# Animals in the Anthropocene

ANIMAL PUBLICS

Melissa Boyde & Fiona Probyn-Rapsey, Series Editors

---

*Engaging with animals: interpretations of a shared existence*
Ed. Georgette Leah Burns & Mandy Paterson

*Animal death*
Ed. Jay Johnston & Fiona Probyn-Rapsey

*Cane toads: a tale of sugar, politics and flawed science*
Nigel Turvey

# Animals in the Anthropocene

Critical perspectives on non-human futures

Edited by the Human Animal Research
Network Editorial Collective

SYDNEY UNIVERSITY PRESS

First published by Sydney University Press
© Individual contributors 2015
© Sydney University Press 2015

Sydney University Press
Fisher Library F03
University of Sydney NSW 2006
AUSTRALIA
Email: sup.info@sydney.edu.au
sydney.edu.au/sup

**National Library of Australia Cataloguing-in-Publication Data**

| | |
|---|---|
| Title: | Animals in the Anthropocene : critical perspectives on non-human futures / edited by the Human Animal Research Network Editorial Collective. |
| ISBN: | 9781743324394 (paperback) |
| | 9781743324400 (ebook: epub) |
| | 9781743324417 (ebook: mobipocket) |
| Series | Animal Publics. |
| Notes: | Includes bibliographical references and index. |
| Subjects: | Animals and civilization. |
| | Nature--Effect of human beings on. |
| | Human-animal relationships. |
| | Ecology. |
| Other Creators/ Contributors: | Human Animal Research Network Editorial Collective, editor. |
| Dewey Number: | 590 |

Cover image: video still from *New World Order* (2013) by Hayden Fowler
Cover design by Miguel Yamin

# Contents

# Introduction

In 2000, Paul J Crutzen and Eugene F Stoermer offered the startling suggestion that the global impacts of human activities during the last 300 years, including human population growth, fossil fuel consumption and greenhouse gas emissions, are so significant in scale that they might constitute a new geological epoch. Importantly, they argued that this new epoch must be understood as being fundamentally shaped by humans:

> considering these and many other major and still growing impacts of human activities on earth and atmosphere, and at all, including global, scales, it seems to us more appropriate to emphasise the central role of mankind in geology and ecology by proposing to use the term "anthropocene" for the current geological epoch'. (Crutzen & Stoermer 2000, 17; see also Crutzen 2002)

The proposed term has not as yet gained official status within the field of geology; however as a frame for understanding a period of geological time marked by the significant impact of human activity on the

HARN Editorial Collective (2015). Introduction. In Human Animal Research Network Editorial Collective (Eds). *Animals in the Anthropocene: critical perspectives on non-human futures.* Sydney: Sydney University Press.

planet, 'Anthropocene' has extraordinary potential. It is a unique term simultaneously oriented to the past, present and future: as well as delineating the thresholds of past activity that have formed our present, it accommodates the 'built in future change' that is 'currently unknowable' (Zalasiewicz et al. 2010). Jan Zalasiewicz, Mark Williams, Will Steffen and Paul Crutzen further stress the importance of thinking about the Anthropocene in terms of its intertwining 'forces', conceived as human and natural, where 'the fate of one determines the fate of the other' (2231). Beyond the significance that it holds for geologists, they also recognise that the Anthropocene 'has the capacity to become the most politicised unit, by far, of the Geological Time Scale' (2231). The high stakes of this politicisation of the history of the Earth calls out for rigorous analysis of the term's usefulness, its effects, its tensions, its limitations, and its potential.

Much of the focus of discussion on the Anthropocene has centred upon anthropogenic global warming and climate change and the urgency of political and social responses to this problem. However, we argue that there is an equally important challenge in thinking about our relationships with non-human animals. It is worth noting that in Crutzen and Stoermer's original formulation three sites of human–animal interaction – the growth in global cattle populations, species extinction, and the expansion of industrialised fishing – were highlighted by the authors as significant markers of human impacts within the Anthropocene (Crutzen & Stoermer 2000, 17). Attending to human power over non-human life is an urgent task, and solutions to pressing environmental problems (such as climate change) can only be sought by recognising their interdependence with how we relate to animals (for example, in reducing the impact of industrial animal utilisation for food production). It is from the perspective of 'the animal question' – asking how best to think and to live with animals – that this book seeks to interrogate the Anthropocene as a concept, discourse, and state of affairs. What conceptual frameworks might make use of, critique and expand on the term's usefulness? How might the Anthropocene be 'politicised' and what 'feedback effects' might the term produce? How might the Anthropocene enable, provoke and frame discussions of multispecies responsibility and justice? How might the term intersect with, and respond to, global inequalities across species, class, gender, race, ability, sexuality and other divides? Does the concept of the Anthropocene disrupt the persistent notion that humans transcend planetary life, or does the entire discourse re-centre humans yet again?

The application of the term 'Anthropocene' brings with it certain risks and problematic tendencies about the 'Anthropos' ('human' or 'man' in Greek) it names. One danger is that anthropogenic destruction of natural systems and domination and extinction of other animals are taken as natural and inevitable results of the development of *Homo sapiens* as a species. Our impacts become the inevitable outcome of humankind's 'super-natural' nature – events which cannot be overturned but might only be mitigated through planetary management and geo-engineering (Crist 2013). Another related danger is reflected in the irony of its naming: at the same time as the Anthropocene definitively reveals humankind's inescapable dependency on ecological processes that support life on Earth, it also highlights the extent to which human agency has transcended and has threatened such processes (Dibley 2012). Humanity is elevated to '*The God species*' (Lynas 2011), a force more powerful than nature, capable of transforming Earth itself. As such, the apocalyptic tenor that has long sounded in environmentalists' warnings becomes here more intense and complex – 'we' are both problem *and* solution.

The discourse of the Anthropocene manifestly perpetuates and deepens a dominant Western hierarchy, at the same time that it *scrambles* and troubles the dualisms on which it rests. It is on the side of the scramblers that this book finds company. The multiple voices in this book, representing a range of disciplines all concerned with the contemporary question of the animal, seek to complicate the discourse of the Anthropocene: to open up further spaces for questioning the uneven effects of climate change and environmental destruction on different human and non-human groups; to engage with the social and political dimensions of the Anthropocene; to challenge its naturalisation while facing up to the irremediable changes that it signifies.

The need to decentre the human in discourses on the Anthropocene is pressing. The Anthropos that gives its name to this era inherits much from the history of philosophical and theological concepts of humanity that place the human at the centre of life and history. In their deployment of the concept of the Anthropocene, scientists and humanists commonly revert to traditional tropes, narratives, and concepts of the human. Judaeo–Christian philosophical anthropology has been foundational in Western culture, whether the *imago Dei* that places humanity above the animals and gives to this 'man' the sovereign right of dominion, or the fall of original sin that separates knowing humanity from the innocence of nature. There are of course noteworthy minority strands – the figure of St Francis is significant – and recent ecocritical readings of the Bible have, for example, challenged the command to subdue the Earth often interpreted from Genesis, emphasising instead stewardship, democratic creatureliness, or other animal proximities (Habel & Wurst 2000). However, the dominant Christian axioms have remained influential in so-called secular modernity, and many have held them to be significantly responsible for the West's environmental destruction (White 1967).

Continuities can certainly be traced from these anthropocentric theologemes to the philosophemes of modernity, which as Jacques Derrida (2008) argues share a 'carnophallogocentric' structure that privileges a carnivorous, speaking 'man'. Twentieth-century philosophy attempted to historicise and deconstruct this ontotheology of God and 'man' – as Friedrich Nietzsche (1974, 167) put it, to vanquish the shadows of a never quite dead enough God. With the rise of animal studies the *species* dimension of this humanism has begun to be thoroughly elucidated and analysed. As the Anthropocene names the human as a *species* – one among many – the term is potentially useful for displacing human mastery in favour of human–animal relationality. Indeed, many of the chapters in this book re-appropriate the term as a key driver for this reorientation towards relationality and away from presumptions of human mastery and separation that are embedded in anthropocentric thinking.

As animal studies scholars have consistently shown, anthropocentrism carries with it an assumption of human ascendancy across a range of criteria – for example, intelligence, creativity, freedom, morality, reasoning – all of which posit agency as belonging only to humans. And as Florence Chiew shows in 'The paradox of self-reference' (Chapter 1), such problematic humanist assumptions too often inform the use of these concepts in scientific and sociological literature about the Anthropocene. Chiew applies insights from sociology and systems theory to help rethink *agency* and *intervention* – crucial concepts in the discourse of the Anthropocene. What does it mean to act, to intervene? Analysing the contemporary philosophical and scientific discussion of the Anthropocene, she points out not only its human-centredness, but in particular its self-referentiality, naming the 'human' as the source of both destruction and restoration, both culpable and curative, both subjected to natural processes and subjecting nature to our own designs. She pays careful attention to the foundational sociology of Émile Durkheim, and the transformational systems theory of Niklas Luhmann and Gregory Bateson, in order to elaborate our understanding of human intervention and agency, and to work towards its reconceptualisation in terms of entanglement and relationality.

The implication of relationality and recognition are also explored in Ben Dibley's chapter (Chapter 2), 'Anthropocene: the enigma of "the geomorphic fold"'. Dibley describes the Anthropocene as a 'difficult dance' that must embrace a 'strange paradox' that recognises 'a distributed humanity, which, with its prosthetics and animals, now composes a geological agent; while taking responsibility for the [legacy of the] other'. Dibley's chapter surveys how the term has provoked reconceptualisations of time, human agency and responsibility as unknowable and yet 'written in stone'. Dibley and Chiew's rethinking of the Anthropocene contrasts with, but also builds on, contemporary scientific debates about the Anthropocene as geological time.

Scientific discourse emphasises the Great Acceleration of anthropogenic environmental changes since the 1950s, amid the contentious 'golden spike' debate about the technicalities of the recognition of the Anthropocene as a new geological epoch (Steffen, Grinevald et al. 2011). Many proposed solutions treat the symptoms (such as through geo-engineering or solar radiation management) instead of removing or reducing the anthropogenic pressures on the environment (Steffen, Gordon et al. 2011). Social sciences and humanities disciplines contribute to this debate by shifting the boundaries of inquiry to consider not only effects and mitigations but also the materiality of cultural life and its resultant 'causes'. For example, Smith and Zeder (2013) and other archaeologists (Balter 2013; Rick et al. 2013), point to animal and plant domestication in the early Holocene as the pivotal point in human ability to modify ecosystems. In this view, domestication of human, animal and plant communities initiated a series of ecological transformations that has contributed directly to the Great Acceleration. The various dates suggested for the onset of the Anthropocene could be used to define successive phases within the Holocene/Anthropocene epoch. Placing the onset of the Anthropocene at the Pleistocene–Holocene boundary (c. 11,000 to 9000 years ago) not only broadens the scope of inquiry to encompass the entirety of human/animal engineering of Earth's ecosystem, but also allows us to build greater understanding of the role of human and animal societies of the past, while recognising the diachronous character of anthropogenic activities around the world (Price et al. 2011; Gibbard & Walker 2013).

In Chapter 3, 'Cycles of anthropocenic interdependencies on the island of Cyprus', Agata Mrva-Montoya highlights the different time scales involved in human engineering of animal life. Archaeological records show that the ecology of the island of Cyprus has been shaped by human management for more than 10,000 years, including introductions (both accidental and deliberate), and extinctions (also both accidental and deliberate). Mrva-Montoya points out that the 'cycle of introduction, feralisation, naturalisation, control and extinction continues to the present' and that much of this cycle is driven by dichotomies of ecological belonging embedded in the terms feral/wild, native/exotic. These terms still carry explanatory power even though the archaeological and historical evidence shows them to be obsolete at best. Taking such a long view of human folly and the shaping of the island by human hands shows how fraught and how unpredictable 'human management' of life is, and how quickly values can change. Such a long-term view must temper, according to Mrva-Montoya, any contemporary claims to be able to 'mitigate' or take 'responsibility for anthropocenic damage in general'. As her work shows, if there is one consistency in a field of unpredictability, it is that in the last few thousand years, the island is 'littered with examples of fixing one problem by introducing another'. That is surely a defining feature of the term 'Anthropocene' that, suggestively, if not seductively, names 'human' interventions as both problem and solution simultaneously.

The sense of urgency about human–non-human interactions in the Anthropocene calls for a reassessment of all of our relationships with domestic and 'wild' animals, those species that have co-constitutively evolved alongside humans in specific political communities and designated spaces such as the 'wilderness'. Alongside environmental crisis, the Anthropocene brings a new wave of projects aimed at controlling populations of wild animals. In particular, nationalist discourses shape conceptualisations of nature and culture, and the environmental politics that emerges therefrom. In Chapter 4, Adrian Franklin discusses the impact of the longer time frame of the Anthropocene suggested by some researchers on the perception of humans as 'planet changers by nature', questioning the myth of 'pre-industrial peoples as a "model species"'. Franklin compares the different responses to feral and introduced animals in Australia and Britain, linking the levels of tolerance and acceptance to ideas of nation. In Australia various agencies advocate countless forms of eradication in the name of preservation of indigenous species, while Britain's tolerance and even fondness for introduced species attests to its 'cosmopolitan' attitudes and reflects, most probably, a sustained history of species introductions over a long period of time. Franklin points out (following Nicholas Smith) that this xenophobic discourse functions as a way of indigenising the settler self and naturalising national boundaries. Franklin shows that 'ecosystem is used inconsistently, a smokescreen for self-evidently social processes'. Both concepts of ecosystem and landscape 'arose in the Anthropocene to help us make sense of changing/changed worlds and guide action'.

If the 'human species' is the agent of its own decline, it is then also clear from postcolonial, first nations and ecofeminist perspectives that responsibility for and even response-ability towards anthropogenic decline is not equally shared, nor its effects equally distributed across the landscapes of the Anthropocene. Vanessa Barbay (Chapter 5) demonstrates that an Indigenous concept of Country contrasts with European settler notions of 'landscape' (and the 'environment'). The latter is framed by distance, containment and separation of self and Others (human and non-human). In Australia, the collision of landscape with Country shapes different ways of belonging that are difficult, if not impossible, to reconcile. Landscapes gain expression not through embodied relations to place and so the term lends itself to an ideology of resource management. As Irene Watson points out, the failure to recognise Country is inextricably linked with the denial of 'pre-existing Aboriginal laws that have lived in this land from the beginning' (Watson 2002, para 3). The concept of 'Country' is an alternative to what Val Plumwood calls the *hyperseparation* that marks Western accounts of living in the landscape or the 'environment.' The contrasts between Indigenous knowledges and Western separations are great, and are perhaps noticed more now than ever before when relational approaches to human animal life gain wider attention and applicability.

Such relational approaches have been developed by a variety of feminist scholars within animal studies to explain human and animal interconnection (for example, Adams 2007; Haraway 2008; Oliver 2009; Adams & Gruen 2014). Donna Haraway, for example, draws attention to the way in which humans and animal coshape each other: 'we are in a knot of species coshaping one another in layers of reciprocating complexity all the way down' (2008, 42). Such coshapings are a claim for mutual recognition; yet this is explicitly obscured by the self-naming 'Anthropos'. That is, specifically sexed and raced bodies have played different roles in the intensification of 'human' impacts. Moreover, 'human' is a term that, as Rosi Braidotti explains, presumes an 'allegiance' that is, for feminists, 'at best negotiable' (2009, 531). Instead feminist work has been orientated towards woman, the feminine and feminised, the 'more than human' (Whatmore 2002), and the 'never been human' (Haraway 2008). Such 'disloyalties' to human civilisation (to appropriate Adrienne Rich's term) have been celebrated in numerous (broadly) ecofeminist works dating back decades (see, for example, Wright 1968; Daly 1978) whose proposals on relationality are gaining new relevance and applicability (see, for example, Lori Gruen 2015). In Chapter 6, Madeleine Boyd draws explicitly from Karen Barad's agential realism (Barad 2007) and new materialist feminism to argue for an interconnection between humans and horses that is founded upon relationality; a framework that allows Boyd to both critically understand the horseracing industry as 'a system of intra-actions-that-matter between significant bodies, material limitations' and suggest that there is scope to imagine a renewed 'horse and human dyad.' Through her interactive artwork ('A game of horseshoes for the ineffectual martyr') Boyd draws attention to the historical interspecies dyad that formed part of the narrative for the development of human cultures and institutions; a relationship which in the contemporary context Boyd argues has been 'wrenched apart' to make way for the commodification associated with the horseracing industry. In this sense Boyd argues that the 'ancient social contract' with horses has been broken by over-breeding, racing and wastage practices. In relation to the challenges posed by the Anthropocene, Boyd urges us to 'embrace becoming *Homo sapiens relationata* while allowing *Homo destructus* to fade'.

In Chapter 7, Daniel Kirjner argues that 'environmental transformations that characterise the Anthropocene cannot be considered apart from norms of masculinity'. Kirjner explores the impact of human activities on non-human animal life through the lens of domination, examining specifically the way in which industrialised food production directly exerts large-scale harm on animals used as food, and simultaneously produces environmental side effects for other animals who are impacted by this production: 'the development of one species for animal agriculture triggers the reduction and extinction of many others'. Drawing on ecofeminist philosophy, Kirjner observes that human use of animals produces a distinctively patriarchal form of domination, and is inherent within an overarching culture that celebrates violence and aggression. Citing Carol J Adams' classic work, *The sexual politics of meat*, Kirjner reinforces that human violence within the Anthropocene rests upon production of an 'absent referent', whereby violence is veiled through metaphor: 'pig' becomes 'pork'; 'tree' becomes '2x4'. In understanding the specifics of this violence, Kirjner draws attention to a violence of predation, arguing that this is the central and defining characteristic of (human) male-centric domination within the Anthropocene.

Following on from Kirjner's discussion of the politics of meat consumption, Simone Dennis and Alison Witchard (Chapter 8) explore the ontological implications and potential biopolitical consequences of the development of in-vitro meat. At the cutting edge of synthetic biology, the production of this 'artificial' form of fleshy nutrition does not rely on industrial agricultural systems, rituals of killing, or the existence of a whole animal. Meat is a form of capital that materialises social and species hierarchies. The gradual commodification of the processes of meat production and consumption are widely held to reflect the techno-social alienation of humankind from the rest of creation. The pursuit of 'good' in-vitro meat has the potential to disrupt and unsettle these arrangements. Because meat consumption embodies deeply embedded socio-moral assumptions, the production of in-vitro flesh also affords us the opportunity for the obliteration of carnal hierarchies – flesh produced from human cells could theoretically end up on the table for human consumption. Even as the molecular fabrication of meat intends the liberation of other species it could completely reframe how we situate ourselves in the world. To take this to a post-Fordist extreme – if we begin to eat and to consume ourselves then do we transcend or reaffirm our animality? Test-tube meat could initiate and materialise new forms of labour and production, and, thereby, a socio-moral order in which human domination is not pregiven and all flesh is rendered equally open to commodification. Dennis and Witchard suggest that within the potential for the erasure of 'species' thinking we can see traces of the birth of the Anthropocene – where longstanding categories and binaries break down and force us to reconceptualise what it means to be a 'being'. Could the animals to whose suffering many have turned a blind eye then become our *someones* rather than our some-things? Could, as Carol Adams and Tom Tyler ask, livestock also be our companion species (Adams & Tyler 2006, 126)? And if this happened, what would it mean for the future of our planet and all of those with whom we share it? Thus, tied to the new ethico-political projects that are implied by the conceptualisation of the Anthropocene is a need to imagine new futures, including a futurity with non-human animals.

Gwendolyn Blue argues for rethinking the notion of the 'public' in posthumanist terms in Chapter 9. Blue calls for 'an understanding of the public as a multispecies rather than a strictly human accomplishment' in the context of Anthropocene, engaging with the theories of American pragmatism (John Dewey) and feminist science studies (Karen Barad and Donna Haraway). Drawing on Barad's theory of intra-action, the author describes publics 'as phenomena, as the material-discursive entanglements that emerge in tandem with technological, scientific, cultural and geological transformations'. Blue argues that the abilities for inquiry and communication are two important elements of the participation of the non-human in publics in which they are already inherently included through the implications of the public's actions. A multispecies public moves away from anthropocentrism in order to examine how the public contributes to the ways in which differences, boundaries and distinctions are made in the world. A multispecies public explores the ways in which conceptual binaries are constructed and enacted, and raises questions about response-ability in the context of the transformations in the Anthropocene.

As anthropogenic change affects the more-than-human world in innumerable ways, many argue that we must accept responsibility for the damage we have caused, and the debt we owe to non-human species. If we extend the status of companions to other species such as those who are, in Val Plumwood's terms, 'active presences and ecological collaborators in our lives' (2002, 195), we reach out with compassion and empathy not just to the animals that share with us our domestic spaces, but those with whom naturecultures far and wide are constituted. In Chapter 10, Richie Nimmo analyses the paradox in the concept of the Anthropocene; a tension between its powerful rejection of notions of human transcendence of the natural world and its implicit but unmistakable call for human responsibility on a scale that presupposes such transcendence. Diagnosing this as a tension between contrasting conceptions of human agency, the one as ultimately sovereign and determinative, the other as at best partial, contingent and entangled with various non-human agencies, Nimmo traces the same paradox through the tension between critical and posthumanist currents in human–animal studies. Nimmo explores this tension through the recent and rapid worldwide decline in honeybee populations known as colony collapse disorder (CCD), exactly the sort of socio-ecological crisis we should expect to see more frequently in the Anthropocene. A critical animal studies approach to CCD is evaluated, with honeybees viewed as a kind of 'livestock' and commercial beekeeping or apiculture understood as part of an 'animal-industrial complex'. Nimmo then articulates a posthumanist approach, which he suggests is better equipped to acknowledge the specificities of honeybees and the nuances of human–apian relations, before reckoning the implications of these alternative accounts of CCD back into the paradoxical notions of human agency and responsibility at the heart of the Anthropocene. Again, the question of living with other animals in a sphere of response-ability and community is at the forefront of reorientating human–animal relations in the Anthropocene.

Krithika Srinivasan (Chapter 11) takes us to the streets of Bengaluru, India, to examine and unpick the biopolitics of controlling stray dog populations. Drawing on Michel Foucault, she describes how the pursuit of human wellbeing through practices of dog control intersects with changes in the conceptualisation and practice of animal welfare. In doing so, Srinivasan reveals the central opposition and normative paradox embedded in the Anthropocene. Given our power over life and capacity to accept responsibility, how should we seek to live so as to protect the wellbeing of non-human others while also safeguarding human interests? In particular she reflects upon the entrenchment of norms about the sanctity of human interests in what she describes as the welfare episteme, and how this structures the manner in which street dog wellbeing is understood and pursued through vaccination and neutering programs. Central to this are discourses and practices of benevolence within which it becomes unthinkable that street dogs could thrive or flourish unless they are under human care. Under the welfare episteme, as with other forms of governmentality, the boundaries between individuals and populations, harm and care, and welfare and control are blurred to permit and valorise acts of violence on others. As humans increasingly seek to intervene in the lives of animals for their own good, Srinivasan asks us to consider whether the forms of social change that will emerge in response to the Anthropocene will also necessarily be marked by the exercise of similar forms of non-benign power.

Blue, Nimmo and Srinivasan's chapters are working largely against conventional political theory in which non-human animals are not treated as belonging to, or participating within, the civil political sphere. This exclusion is famously pronounced in Aristotle's *Politics*, where the philosopher states unambiguously that 'man is by nature a political animal' (Aristotle, *Pol*.I, 1253a). For Aristotle a human capacity for speech and justice differentiates this species from others: 'it is a characteristic of man that he alone has any sense of good and evil, of just and unjust, and the like, and the association of living beings who have this sense makes a family and a state' (*Pol*.I, 1253a). In some respect political philosophy has failed to move forward from this flawed assumption. It is true that work in moral philosophy has sought to challenge the speciesist norms which supposedly justify human indifference to animal status, welfare and rights (Singer 1975; Regan 1983; Nussbaum 2006; Francione 2007). However this work has failed to dent political philosophy, which has largely, it would appear, remained impervious to thinking about animals, resistant to understanding human treatment of animals through the lens of politics and power, and often not understood animals as potentially subjects of political agency.

Against this historical bias, a range of contemporary liberal political theorists are seeking ways to not only include consideration of animals as worthy of moral consideration and rights, but as belonging in some way to the political sphere as participants and subjects of justice (Garner 2005; 2013; Kim 2007; Cochrane 2010; Donaldson & Kymlicka 2011; O'Sullivan 2011). At the same time, other thinkers are drawing from the critical and continental philosophical tradition to rethink the exclusion of animals from the political sphere, and consider the ways in which human violence towards non-human animals is interconnected with human violence towards other humans and linked to global systems of production (Benton 1993; Shukin 2009; Chrulew 2012; Wolfe 2012; Pugliese 2013). This work also creates unprecedented opportunities to understand the nature of human domination of animals as a political problem, extending pioneering work done by thinkers such as Val Plumwood (1993; 2002) and Barbara Noske (1997). These new directions create opportunities to critique large-scale political arrangements, forms of systemic violence and political subjectivities that harm non-human life, and simultaneously, to imagine new forms of relationality that may produce shared futures with animals.

Attempts to 'rethink the political' in the Anthropocene should also remind us that there remains an ongoing challenge in how we might rethink 'the social', and understand human relationships with animals as worthy of sociological exploration. There is a growing field of work in the area of animals and society (Arluke & Sanders 1996; Peggs 2012; Taylor 2013) and emerging analysis of production and biotechnology (Twine 2010), the dynamics of animal slaughter and food production (Pachirat 2011), and the dynamics of companionship and domestication (Haraway 2008), often intersecting with interdisciplinary fields, such as population health (Blue & Rock, 2011). If we are to imagine, like Jennifer Wolch (1998), an embracing zoopolis that can sustain close interrelationships between humans and animals, we must then surely dream of new forms of connectivity, different social arrangements, and new forms for the political subject within this new world (Gibson-Graham & Roelvink 2010).

The final chapter, 'Wild elephants as actors in the Anthropocene' by Michael Hathaway (Chapter 12) explores the concept of non-human agency with respect to human–elephant relationships in Yunnan Province, Southwest China. Hathaway spent a year living with his family in a small village where he conducted fieldwork that explored the complex relationships between rural villagers, a powerful conservation organisation, highway developers, and China's last herds of wild elephants. He recounts stories that show the purposeful agency of these animals, of the mindful behaviour that has thwarted attempts to stop elephants raiding crops, and that has caused major disruptions in the construction of a new highway joining Yunnan's Kunming city to Bangkok. The tensions between herd survival in the face of habitat loss, the livelihoods of rural villagers, and China's technological development are the source of ongoing human–elephant conflict in Yunnan Province. But Hathaway argues that the agential behaviour of elephants – and their status as crop raiders and trespassers – is not a product of 'resistance' against humans. The behaviour that causes conflict is just one element of their rich lives, and is central to fulfilling their need for food and their desire for learning, exploration and experimentation. The elephants' unwitting conflict with humans is concomitant with the fraught relationships that exist between humans and animals in the Anthropocene. However, as his chapter shows, China has made concerted attempts to mediate human–elephant conflict without punitive forms of control. As a result China has the potential to engender more peaceful interactions between humans and elephants than those in Africa, and throughout Asia, where 'rogue' animals are often killed. In the face of a rapidly declining elephant population, forms of advocacy that negotiate conflicts, rather than resolve them through brutality, could pave the way forward for human–animal relationships looking beyond the Anthropocene.

*Animals in the Anthropocene: critical perspectives on non-human futures* looks beyond the current anthropocentric doxa of human dominance over other animals. As such, we find in the term 'Anthropocene' a useful device for drawing attention to the devastations wreaked by anthropocentrism and advancing a relational model for human and non-human life. The effects of human political and economic systems on animals continue to expand and to intensify, in numerous domains and in ways that not only cause suffering and loss but that also produce new forms of life and alter the very nature of species. An insight into new life and altered natures is provided by Hayden Fowler's artwork, *New World Order* (2013) and the accompanying artist's statement, included here as an epilogue to this book. Fowler's work, like all the chapters in this book, shows that assessing the effects of human activity on the planet requires more than just the quantification of ecological impacts towards the categorisation of geological eras. It requires recalibrating our standpoints to recognise and to evaluate a wider range of territories and terrains, full of non-human agents, interests and meanings that are exposed to transformation by this paradoxically immanent and powerful agent of change that gives its name to the Anthropocene.

*HARN Editorial Collective: Madeleine Boyd, Matthew Chrulew, Chris Degeling, Agata Mrva-Montoya, Fiona Probyn-Rapsey, Nikki Savvides & Dinesh Wadiwel*

## Works cited

Adams CJ ([1990] 2010). *The sexual politics of meat: a feminist-vegetarian critical theory*. New York: Continuum.

Adams CJ (2007). *The feminist care tradition in animal ethics: a reader*. New York & Chichester, UK: Columbia University Press.

Adams CJ & Gruen L (2014) *Ecofeminism: feminist intersections with other animals and the earth*. London: Bloomsbury.

Adams CJ & Tyler T (2006). An animal manifesto: gender, identity, and vegan-feminism in the twenty-first century. *Parallax* 1: 120–28.

Aristotle (1952). *Politics*. RM Hutchins (Ed). *The works of Aristotle*. Vol. 2 (pp445–548). Chicago, IL: Encyclopaedia Britannica.

Arluke A & Sanders C (1996). *Regarding animals*. Philadelphia, PA: Temple University Press.

Balter M (2013). Archaeologists say the 'Anthropocene' is here – but it began long ago. *Science* 340(6130): 261–62.

Barad KM (2007). *Meeting the universe halfway: quantum physics and the entanglement of matter and meaning*. Durham, NC: Duke University Press.

Benton T (1993). *Natural relations: ecology, animal rights and social justice*. London: Verso.

Blue G & Rock M (2011). Trans-biopolitics: complexity in interspecies relations. *Health* 15(4): 353–68.

Braidotti R (2009). Animals, anomalies and inorganic others. *PMLA* 124: 526–32.

Chrulew M (2012). Animals in biopolitical theory: between Agamben and Negri. *New Formations* 76: 53–67.

Cochrane A (2010). *An introduction to animals and political theory*. Basingstoke, UK & New York: Palgrave Macmillan.

Crist E (2013). On the poverty of our nomenclature. *Environmental Humanities* 3: 129–47.

Crutzen PJ (2002). Geology of mankind. *Nature* 415: 23.

Crutzen P & Stoermer EF (2000). The 'Anthropocene'. *Global Change Newsletter* 41:17–18. Retrieved on 29 April 2015 from http://www.igbp.net/download/ 18.316f18321323470177580001401/NL41.pdf.

Daly M (1978). *Gyn/ecology: the metaethics of radical feminism*. Boston, MA: Beacon Press.

Derrida J (2008). *The animal that therefore I am*. M-L Mallet (Ed), D Wills (Trans). New York: Fordham University Press.

Dibley B (2012). 'The shape of things to come': seven theses on the Anthropocene and attachment. *Australian Humanities Review* 52: 139–53.

Donaldson S & Kymlicka W (2011). *Zoopolis: a political theory of animal rights*. Oxford, UK: Oxford University Press.

Francione GL (2007). *Animals, property and the law*. Philadelphia, PA: Temple University Press.

Garner R (2005). *The political theory of animal rights*. Manchester, UK: Manchester University Press.

Garner R (2013). *A theory of justice for animals*. New York: Oxford University Press.

Gibbard PL & Walker MJC (2013). The term 'Anthropocene' in the context of formal geological classification. In C Waters, J Zalasiewicz, M Williams, M Ellis & A Snelling (Eds). *A stratigraphic basis for the Anthropocene* (p395). London: Geological Society, Special Publications. doi 10.1144/SP395.1.

Gibson-Graham JK & Roelvink G (2010). An economic ethics for the Anthropocene. *Antipode* 41: 320–46.

Gruen, L (2015). *Entangled empathy*. New York: Lantern Books.

Habel NC & Wurst S (Eds) (2000). *The earth story in Genesis*. The Earth Bible, 2. Sheffield, UK: Sheffield Academic Press.

Haraway D (2008). *When species meet*. Minneapolis, MN: University of Minnesota Press.

Kim C (2007). Multiculturalism goes imperial: immigrants, animals and the suppression of moral dialogue. *Du Bois Review: Social Science and Research on Race* 4(1): 233–49.

Lynas M (2011). *The God species: saving the planet in the age of humans.* Washington, DC: National Geographic.

Nietzsche F (1974). *The gay science, with a prelude in rhymes and an appendix of songs.* W Kaufman (Trans). New York: Vintage Books.

Noske B (1997). *Beyond boundaries: humans and animals.* Montreal, QC: Black Rose Books.

Nussbaum MC (2006). *Frontiers of justice: disability, nationality, species membership.* Cambridge, MA: Harvard University Press.

O'Sullivan S (2011). *Animals equality and democracy.* Basingstoke, UK & New York: Palgrave Macmillan.

Oliver K (2009). *Animal lessons.* New York: Columbia University Press.

Pachirat T (2011). *Every twelve seconds: industrialized slaughter and the politics of sight.* New Haven, CT: Yale University Press.

Peggs K (2012). *Animals and sociology.* Basingstoke, UK & New York: Palgrave Macmillan.

Plumwood, V (1993). *Feminism and the mastery of nature.* London & New York: Routledge.

Plumwood V (2002). *Environmental culture: the ecological crisis of reason.* London & New York: Routledge.

Price SJ, Ford JR, Cooper AH & Neal C (2011). Humans as major geological and geomorphological agents in the Anthropocene: the significance of artificial ground in Great Britain. *Philosophical Transactions of the Royal Society A: Mathematical, Physical and Engineering Sciences* 369(1938): 1056–84.

Pugliese J (2013). *State violence and the execution of law: biopolitical caesurae of torture, black sites, drones.* Abingdon, UK: Routledge.

Regan T (1983). *The case for animal rights.* Berkeley, CA: University of California Press.

Rick TC, Kirch PV, Erlandson JM & Fitzpatrick SM (2013). Archaeology, deep history, and the human transformation of island ecosystems. *Anthropocene* 4: 33–45.

Shukin N (2009). *Animal capital.* Minneapolis, MN: University of Minnesota Press.

Singer P (1975). *Animal liberation.* London: Jonathan Cape.

Smith BD & Zeder MA (2013). The onset of the Anthropocene. *Anthropocene* 4: 8–13.

Steffen W, Gordon L, Molina M, Ramanathan V, Rockström J, Scheffer M, Schellnhuber HJ, Svedin U, Persson A, Deutsch L, Zalasiewicz J, Williams M,

Richardson K, Crumley C, Crutzen P & Folke C (2011). The Anthropocene: from global change to planetary stewardship. *AMBIO: A Journal of the Human Environment* 40(7): 739–61.

Steffen W, Grinevald J, Crutzen P & McNeill J (2011). The Anthropocene: conceptual and historical perspectives. *Philosophical Transactions of the Royal Society A: Mathematical, Physical and Engineering Sciences* 369(1938): 842–67.

Taylor, N (2013). *Humans, animals and society: an introduction to human–animal studies.* New York: Lantern Books.

Twine R (2010). *Animals as biotechnology: ethics, sustainability and critical animal studies.* London: Routledge/Earthscan.

Watson, I (2002). Aboriginal laws and the sovereignty of Terra Nullius. *Borderlands ejournal*, 1(2). Retrieved on 2 March 2015 from www.borderlands.net.au/vol1no2_2002/watson_laws.html.

Whatmore S (2002) *Hybrid geographies*. London: Sage.

White L Jr (1967). The historical roots of our ecologic crisis. *Science* 155(3767): 1203–7.

Wolch J (1998). Zoopolis. In J Wolch & J Emel (Eds). *Animal geographies: place, politics and identity in the nature-culture borderlands* (pp119–38). London: Verso.

Wolfe C (2012). *Before the law: human and other animals in a biopolitical frame.* Chicago, IL: Chicago University Press.

Wright J (1968). Conservation as a concept. *Quadrant* 12(1): 29–33.

Zalasiewicz J, Williams M, Steffen, W & Crutzen P (2010). The new world of the Anthropocene. *Environmental Science & Technology* 44(7): 2228–31.

# 1

# The paradox of self-reference: sociological reflections on agency and intervention in the Anthropocene

*Florence Chiew*

The present exigency of ecological crises such as climate change, decline in global biodiversity, and overpopulation has heightened debates about the role of the human species in shaping the future and sustainability of the terrestrial environment. From philosophical and socio-political perspectives to historical and scientific analyses, a miscellaneous body of literature on the predicament of anthropogenic effects is quickly converging on the problem of conceptualising human culpability vis-à-vis the management of environmental risks. This is particularly evident in the surge of enthusiasm for ideas about the 'Anthropocene', a proposed new unit of geological time to evaluate the extent of human influence on the depletion of Earth's natural resources (Crutzen 2002).

Introduced in the relatively young field of Earth system science, the notion of the Anthropocene was first championed by Dutch atmospheric chemist, Paul Crutzen, who was awarded the 1995 Nobel Prize for his research on ozone-depleting chemicals. In a 2002 piece in *Nature*, tellingly titled 'Geology of Mankind', Crutzen defines the spatio-temporal parameters of the Anthropocene in this manner:

Chiew F (2015). The paradox of self-reference: sociological reflections on agency and intervention in the Anthropocene. In Human Animal Research Network Editorial Collective (Eds). *Animals in the Anthropocene: critical perspectives on non-human futures*. Sydney: Sydney University Press.

For the past three centuries, the effects of humans on the global environment have escalated. Because of these anthropogenic emissions of carbon dioxide, global climate may depart significantly from natural behaviour for many millennia to come. It seems appropriate to assign the term 'Anthropocene' to the present, in many ways human-dominated, geological epoch, supplementing the Holocene – the warm period of the past 10–12 millennia. (Crutzen 2002, 23)

According to Crutzen, the rapid growth of anthropogenic effects has a very brief history, beginning with the Industrial Revolution of the late 18th century, especially as that period coincided with the invention of the Watt steam engine and subsequently, with records of rising levels of carbon dioxide in the atmosphere (Crutzen 2002, 23).

Crutzen's claim has been explored by geologists Jan Zalasiewicz, Mark Williams, Alan Haywood and Michael Ellis in 'The Anthropocene: a new epoch of geological time?' (2011). The central argument in this study is that the development of the cityscape has enabled the 'right' conditions for the emergence of the Anthropocene. For instance, Zalasiewicz et al. (2011) contend that while commonly cited examples of human-induced stresses on the environment include the extinction of megafauna and the rise of agriculture and deforestation leading to elevated carbon dioxide levels in the atmosphere, there is nevertheless a more salient cause underpinning these problems, namely, urbanisation. As a special phenomenon, Zalasiewicz et al. (2011, 836) argue that urbanisation deserves to be studied in its own right since '[c]ities . . . are now the most visible expression of human influence on the planet'. In another landmark paper on this topic, 'The New World of the Anthropocene' (2010), Zalasiewicz, Williams, Steffen and Crutzen remark that the manmade urban landscape has become a 'geologically unique' phenomenon that is refashioning the conditions and possibilities of life on Earth: '[w]e are . . . living within it, and are able to observe landscape, assess living biodiversity, measure atmospheric composition and sea temperature, gauge ice thickness and sea level height' (Zalasiewicz et al. 2010, 2230). These scientists do not touch on the deeper implications of this unusual geological phenomenon, but what they are pointing to is the paradox inherent in the notion of the Anthropocene, that human activity is both a marker of destruction and of achievement.

Given its relevance to the enduring debates over the nature–society relation, it is not surprising that the term 'Anthropocene' is fast entering the vernacular of the humanities and social sciences. As a few scholars have pointed out, the proposal of the Anthropocene as a human-authored geological event evokes a curious paradox (for example, Latour 2010; Alberts 2011; Dibley 2012; Sayre 2012). The aim of advancing Anthropocene studies is presumably to raise an environmental consciousness and to encourage more effective conservation measures toward ecological crises (Zalasiewicz et al. 2010, 2231). However, by emphasising the unprecedented dominance of human activity in scarring the landscape, the Anthropocene is by definition wedded to a position of human exceptionalism insofar as it distinguishes human sociality as the autonomous source of corrective for environmental devastation. Stated differently, the question of what constitutes intervention and agency in theories of the Anthropocene is not only human-centred, but also paradoxically self-referential. As Ben Dibley puts it, the Anthropocene marks 'the advent of a geological era of humanity's own making' (Dibley 2012, 139).

This seemingly tautological manner of conceptualising human agency evokes something peculiar, for the very identification of a problem as that which is human-created, yet requires human responsibility to resolve would infer that the source of blame is also the force of intervention. If the point of intervention emerges from the same agential capacity for erroneous, wayward behaviour, it becomes clear that this puzzle of self-reference questions the nature of the difference between the corrective and the culpable. At issue here is the problem of delineating the identity of the human as a discrete entity upon which blame or remedy can be readily attributed.

Consider the following examples of how responsibility is framed in the context of environmental crises. For philosopher Paul Alberts when 'the human [is] recalibrated as a geological agent' (Alberts 2011, 5):

> The Anthropocene is a type of limit recognition or emerging limit 'experience.' The conditions for *all* species life are now subject to human decisions. Having accelerated exponentially in force and effect across the last two centuries those human decisions appear today as almost impossible to alter, and increasingly heading towards un-

wanted futures ... we are tying the conditions for all species into a future trajectory that needs to be altered, yet the scale of that situation reaches to the extent of the biosphere. (Alberts 2011, 7, italics in original)

For Alberts, the experience of culpability depends on recognising human existence as a limiting condition for existence in general. The corollary of this view, however, unwittingly lapses into a kind of solipsism because human experience has been cast as something so absolutely autonomous it is radically severed from the experience of the rest of life, positioning humankind outside the object (the conditions for species life) of violation. When figured this way, responsibility becomes a unilateral decision in which duty of care proceeds from the human to other non-human species that would receive, rather than enable it.[1]

Interestingly, if for Alberts the Anthropocene marks the fact that human decisions are failing to sustain the conditions of life for the biosphere as a whole, for historian Dipesh Chakrabarty the problem is the other way around: global warming imperils not the planetary system but the chances of survival for humanity. Tackling the question of human agency from a different angle Chakrabarty forewarns that the history and sustainability of human existence is contingent on, and emerges from, the deep history of the Earth. In this vein, he argues:

The consequences [of the Anthropocene] make sense only if we think of humans as a form of life and look on human history as part of the history of life on this planet. For, ultimately, what the warming of the planet threatens is not the geological planet itself but the very conditions, both biological and geological, on which the survival of human life as developed in the Holocene period depends. (Chakrabarty 2009, 213)

Although Chakrabarty draws on the fundamental insight that humankind is an instantiation of planetary life, he does not pursue the

---

1    Elsewhere, I have explored in greater detail the complexity of theorising responsibility vis-à-vis human–animal relations. See Chiew 2014.

deeper implications of this point. While he rightly acknowledges the complexity of attributing blame to particular groups of human actors or communities, and encourages a reading of responsibility that evokes collective involvement 'aris[ing] from a shared sense of a catastrophe' (Chakrabarty 2009, 222), Chakrabarty's argument leaves the notion of the human unchallenged. As a consequence the distinction between the human and the non-human, or that between the human and ecological processes, is preserved, rather than interrogated anew.

Taken together, both Alberts and Chakrabarty's formulations of human agency presents us with a notion of 'humanity' and 'nature' as a meeting of two separable spheres of life that pre-exist each other. Significantly, the underlying assumption that is operative here, as geographer Nathan Sayre points out, is that '[h]umans, climate, soils, and nonhuman biota are taken as conceptually distinct though empirically interactive; nature is presumptively static or cyclical until proven otherwise; humans are the cause of change, whether progressive or detrimental, incremental or abrupt' (Sayre 2012, 60). In a way, Alberts and Chakrabarty's accounts are mirror images. For Alberts, human actions are so influential and so detrimental in their effects that they threaten the sustainability of planetary life. For Chakrabarty, human actions are endangering not the life of the planet but that of human existence. The former view imbues a sense of autonomy and power to the identity of the human while the latter nurtures a sense of humility and even self-effacement so as to recognise the scale of human life in relation to life in the cosmos. Both readings, however, remain wedded to a self-evident notion of the human, returning us to a dichotomous treatment of the interaction between human and nature that continues to pervade many interventionist projects seeking to pinpoint the source of culpability for environmental degradation.

To be sure, the value of these projects is not in dispute. But what is intriguing for my purposes here is the self-referential paradox that generally goes unacknowledged, or when acknowledged, not taken seriously enough, in attempts to define the human as at once the source of blame and corrective. Accounts of what agency and thus intervention might mean in the Anthropocene capture the fundamental problem of circumscribing the human as simultaneously the subject–author of life *and* an object authored by life. The catch is this: how do we reconcile the contradiction in theorising the human as a part of the natural ecosystem yet seemingly also an independent force acting *on* and, thereby, altering ecosystemic processes? In this self-referential puzzle, if the genesis of humankind cannot be thought apart from the life of the planet as such, what we then mean by *human* agency can no longer be something so obvious that it needs no definition. With a critical eye on this paradox, the guiding question for us is this: what secures intervention or agency as human authored if the identity and the genesis of the 'human' is precisely what is at stake in determining the nature of life in the Anthropocene?

## Self-referentiality: the lineage of social science

The following discussion draws from the insights of sociology, whose disciplinary identity, I would argue, rests on this self-referential puzzle of contemplating what it means to be human. When the French thinker Émile Durkheim established the field of sociology in the late 19th century, his convictions were informed by a shift in methodological focus from the subjective and the particular to the objective and the general, that is, making the study of sociality and human experience a science. Durkheim argued that the extension of the empirical methods of science to apparently 'humanistic' dimensions of life such as morality, politics, and history, would not only enlarge the conventional definition of science and disciplinarity but also, importantly, challenge us to rethink the tenets of human exceptionalism. Crucially, by way of Durkheim's teachings, we have inherited one of the most provocative insights of social science, namely, the inextricability of human experience as both the subject and object of inquiry. Addressing this paradoxical founding of his discipline, Durkheim wisely counselled: 'We cannot visualise existence being called into question, since we see it at the same time as we see our own' (Durkheim [1895] 1982, 63). Social existence, in fact life, does not occupy an external and independent location where the researcher or the scientist, standing apart, observes.

It is this confounding of subject and object of inquiry that, for Durkheim, captures the spirit of social science as it evokes the relational and ecological dimensions of individual biography. In a mode of questioning that resembles and perhaps anticipates current scholarship in Anthropocene studies, social science's foundational concern, a seemingly simple one, is that the specificity of an individual (person, event, or entire species) can only be recognised in the wider frame of existence that sustains it. Nevertheless, the implications of this point are not at all simple. For when we take seriously the entanglement of the particular (human individual or species) and the general (ecology of life) in the Durkheimian sense of part as whole, how the logic of differentiation works is radically reconfigured. Following Durkheim, a given entity obtains its particularity only in its participation as the whole; the identity of this entity, while utterly unique, is not a discrete or locatable unit in the whole. 'It' is an expression of the whole.

This central dilemma of reconciling the notion of an objective social reality with the contingencies of subjective experience is brought home in a telling passage from Durkheim's essay, 'The dualism of human nature and its social conditions'. In it Durkheim contends:

> Although sociology is defined as the science of societies, it cannot, in reality, deal with the human groups that are the immediate object of its investigation without eventually touching on the individual who is the basic element of which these groups are composed. For society can exist only if it penetrates the consciousness of individuals and fashions it in 'its image and resemblance.' We can say, therefore . . . that a great number of our mental states . . . are of social origin. In this case, then, it is the whole that, in a large measure, produces the part; consequently, it is impossible to attempt to explain the whole without explaining the part – without explaining, at least, the part as a result of the whole. (Durkheim [1914] 1973, 149)

We see, here, the fundamental problem of causality that resides at the core of Durkheim's understanding of human agency: that any rendering of individual autonomy is already entangled with the 'whole' of social ecology that animates this individual 'part' into existence. Yet, as Durkheim struggles to clarify, how can the 'whole' not already be manifest in the 'part' if the idiosyncrasies that distinguish individuals of a given social or ecological order are differences that originate from one and the same system of life?

Although Durkheim was working with the individual–social pairing, it must be argued that his complication of the part–whole relation would also apply to the way the human–ecology or human–non-human relation is conceived. That is, if we extend Durkheim's insight to the question of human identity it becomes clear that the difference that secures the uniqueness of humankind is necessarily entangled with the 'non-human' world it is purportedly defined in opposition to. In other words, any rendering of human autonomy is implicated in the wholeness of life that animates the human species into existence. Crucially, this sense of systemic entanglement acknowledges the complexity of locating a 'pure' intervention that is independent of the terrain of ethical quandary. It is here that the meeting of sociology and systems theory opens up a productive space to radically think the locus of this 'system' that both affirms and questions the specificity of the human. On this note, I turn now to the work of German sociologist and systems theorist, Niklas Luhmann, whose commitment to the complexity of ethical intervention as a self-referential problem complicates the relationship between error and corrective as one of human action.

## The world as a self-referential system

For Luhmann, self-referentiality is the raison d'être of the world. In *Social systems*, one of his best-known publications, Luhmann argues that the concept of self-reference and its cognates 'reflection' and 'reflexivity' should be substantially reworked from their 'classical location in human consciousness or in the subject' (Luhmann 1995a, 32) because for him 'self-reference is not a peculiarity of consciousness but comes about in the world of experience' (Luhmann 1995a, 479). In Luhmann's work we witness a profound dilation of the usual scope or scale for thinking about the human self, or the self as human. For what constitutes the self is no ordinary entity confined within a narrow understanding of human experience. Rather, Luhmann is insistent on shifting the frame of discussion from the conventional notions of self and subjectivity to that of the world as system: 'the theme of sociological investigation is not the system of society . . . the theme is the world as a whole' (Luhmann 1989: 7).

Like Durkheim, his predecessor, Luhmann maintains that the specificity and provocation of his discipline rests on the paradoxical notion that the object of analysis, human sociality, is confounded with the inquiring human subject. In other words, the sociologist is an expression of the very sociality she deliberates and tries to makes sense of. For Luhmann, this profound implication of the observer and the observed suggests that every perspective is inevitably its own 'blind spot'. He thus calls attention to the fundamental complication of the self–other, subject–object position, that is, the circularity involved in determining the referent of the 'self'. William Rasch neatly sums up this conundrum that informs much of Luhmann's work: 'How can the self refer to itself without making of itself something other than itself, something that can be referred to, pointed to, as if it were not what was doing the pointing?' (Rasch 2002, 4). In Luhmann's 'world systems theory', this riddle is significant, given that the self is dilated as the world's self-encounter, thus raising the critical question of the status of the observer, or of observation in general. What troubles Luhmann is that most analyses of social interaction rest on a subject-centred humanism or a human-centred subjectivity, that is, the connection between subject and object/ world presumes a notion of relationality as a meeting of two separable domains of life. Against what he sees as a disciplinary blind spot, Luhmann contends that when the paradox of self-reference is opened up as a question of *general systematicity*, we are forced to rethink the nature of inquiry and knowledge beyond their humanist assumptions.

In Luhmann's adage, 'there is no métarécit [metanarrative] because there is no external observer' (Luhmann 1995a, x). A position that bears much resemblance to philosopher Jacques Derrida's famous invocation 'there is no outside of text', Luhmann's systems theory puts him in the company of many who work within the traditions of postmodernism and poststructuralism. Luhmann draws close connections between his notion of self-reference and Derrida's work on parergon or the supplement, noting that deconstruction offers 'the most pertinent description of the self-description of modern society' (Luhmann 1993, 780). For Luhmann, as for Derrida, difference is not a simple separation of two pre-existing entities that then communicate. This means that the differentiating boundary is what both produces and is produced by the system – the world in its entirety. Yet what constitutes this entirety is not an aggregate or sum total of an enclosed space (as the routine reading of totality suggests). Rather, Luhmann's intervention is directed at a different understanding of difference altogether. The genesis of system–environment difference enables knowledge inquiry, for 'difference holds what is differentiated together . . . Differentiation provides the system with systematicity; besides its mere identity (difference from something else), it also acquires a second version of unity (different from itself)' (1995a, 18). For Luhmann, this idea of *unitas multiplex*, or unity in diversity, articulates the double function of the boundary (Luhmann 1995a, 18).

Responding to the question of how binary difference is produced, Luhmann surmises:

> [W]hat is the primary distinction? You could have the distinction finite/infinite, you could have the distinction inside/outside, you could have the distinction being/not-being to start with, and then you can develop all kinds of distinctions in a more or less ontological framework. And I find this fascinating, that there is no exclusive, one right beginning for making a distinction . . . But how are these related? If you engage in one primary distinction, then how do the others come again into your theory or not? This is part of the postmodern idea that there is no right beginning, no beginning in the sense that you *have* to make one certain distinction and you can fully describe the start of your operations. And that's the background against which I always ask, 'what is the unity of a distinction' or 'what do you exclude if you use this distinction and not another one?' (Luhmann 1995c, 30, italics in original)

At first glance, it may be easy to assume that Luhmann is making a point about the relativism and contingency of truth, that our explanations for the way things appear are local to our culture and value system, implying that there is no unchanging or universal standard of justification for why things should be the way they are. Furthermore, such a relativist worldview suggests that any point of origin is contingent, that is, could, in principle, be otherwise. While this may be the conventional reading, I think Luhmann's provocation runs deeper, and insists on a more complicated phenomenon, namely, that there is a logical inconsistency in referring to oneself as an autonomous or impartial arbiter of what is observed because the observed will always 're-enter' or generate a 'reappearance of a difference within the domain of its objects' (Luhmann 1995a, 488).

What fascinates Luhmann is the constitutive nature of exclusions.[2] This is especially telling in his insistence on thinking 'the environment' as forming an essential element of the system of life itself. In *Social systems* Luhmann says:

> The concept of the environment should not be misunderstood as a kind of residual category. Instead, relationship to the environment is *constitutive* in system formation. It does not have merely 'accidental' significance, in comparison with the 'essence' of the system. Nor is the environment significant only for 'preserving' the system, for supplying energy and information. (Luhmann 1995a, 176–77, italics in original)

Importantly, for Luhmann, this inevitable tangle of self-reference (of subject and object, inside and outside) reveals the complexity of circumscribing a perceived error in order to excise it from the 'good' or the desirable. This crucial point bears repeating. For Luhmann, the dilemma that any binary distinction presents is the inherently fraught exercise of attempting to separate the erroneous from the source of in-

---

2   Feminist and quantum physicist Karen Barad makes a similar argument in the context of quantum mechanics. Using the famous two-slit experiment in physics to demonstrate the entanglement of diffraction patterns (wave behaviour) and particles ('bits' of matter), Barad elaborates the quantum insight that the emergence of one entity/part is also an expression of its entire/general frame of reference. What the wave–particle duality paradox shows is the co-constitution of value and measurement, that is, that value or evaluation does not pre-exist the apparatus; it is materialised in and as the specific process of measurement that is being carried out, including the particular exclusions that are enacted. Barad thus makes the important point that the very possibility of drawing any subject/object or observer/observed distinction presents itself as the moment of measurement, evaluation or analysis arises and where certain choices are made to the exclusion of others. There is no unambiguous or inherent distinction between the (human) subject with the agency to observe and to calculate and the object that is presumably non-human and outside the field of agency. See Barad 2007.

tervention as if they were two discrete 'events'. This is, after all, our routine understanding of the relationship between a problem and a solution; causally linked, the problem precedes the solution. Yet, Luhmann's claim of the world as self-referential suggests a more unusual and convoluted sense of interaction and causality. Rather than the linear view of cause and effect, problem and solution, Luhmann would argue that the thing that is understood as a mistake enables the kinds of interventions that are made visible by this purported error. In the context of our discussion that which is deserving of blame in the Anthropocene (human activity) produces the types of actions taken to 'improve' the situation (human activity). This tautology means that what constitutes the culpable is determined as the call for intervention arises and not before it, because in this sense the 'blameworthy' defines the specific intervention that is needed. We will return to this point.

Luhmann is well aware that paradox and tautology offend common sense because they use the distinction between being and non-being, or identity and difference (something 'is' or 'is not') to show that these distinctions 're-enter' their own frames of reference. On this point, he takes his conceptual cue from mathematician George Spencer Brown who says that observing or knowing is in effect an act of discrimination; it relies on, indeed is, an interminable process of demarcating, of cutting up the world in an untold variety of ways. However, the initial 'unmarked' space of the distinction re-enters the observer's frame of reference such that 'all observations have to presuppose *both* sides of the form they use as distinction of "frame". They cannot but operate (live, perceive, think, act, communicate) *within the world.*' (Luhmann 1993, 769, italics in original). Even the frame cannot be conceptualised as a discrete entity or a simple context that merely surrounds the scene of observation for, in order to identify and locate an object, a reference point that holds the differentiating terms must already be available. Luhmann describes this confounding origin of observation as an 'autological conclusion', in that observing 'includes the exclusion of the unobservable, including, moreover, the unobservable par excellence, observation itself, the observer-in-operation' (Luhmann 1995b, 44).

Such a tangled view aims to challenge the conventional assumption that observation relies on a distance between the seer and the seen, and by implication, that sight or insight originates in the human author who controls knowledge of the world. The significance of this point is brought home by the groundbreaking work of anthropologist and polymath, Gregory Bateson. Like Luhmann, Bateson is a systems thinker whose understanding of the nature of observation and communication radically calls into question the human exceptionalism commonly associated with these terms. Author of *Steps to an ecology of mind* ([1972] 2000) and *Mind and nature* ([1979] 2002), Bateson suggests that the intellectual and technological capacities attributed to human uniqueness are more gainfully addressed not by situating humans as exceptional beings above all other species but by demonstrating how human achievement is an instantiation of ecological and evolutionary processes.

In the introduction to *Mind and nature*, Bateson queries: 'What pattern connects the crab to the lobster and the orchid to the primrose and all the four of them to me? And me to you?' (Bateson [1979] 2002, 7). For Bateson, ideation or mental phenomena are not inseparable from the physical environment. Thus the word 'idea', which is etymologically linked to 'form' and 'pattern', takes on an unusually broad frame of reference. Like Luhmann, then, Bateson argues that the experience of what is typically deemed uniquely human is a manifestation of a general ecology of movement, rather than the product of an individual subject or observer. Bateson maintains that insofar as we may say that an entity acts or communicates, this entity cannot be singled out in isolation since individuation is enacted and communicated by a system of relations that is the ecological entanglement of this particular entity. Accordingly, for both Bateson and Luhmann, the question of agency is poorly engaged if it rehearses a conventional representational account of knowledge in which the observer or the knower is self-evidently 'human'. That is, the complexity of the 'intending' subject or author is more productively acknowledged when we can see how human ways of knowing can be generalised to the myriad forms of biospheric interaction observed in plant, insect, and animal worlds. However, the point is not to extend pre-existing human capacities to the apparently non-human world of nature, but to acknowledge, as French philosopher and physician Georges Canguilhem does, the fact that 'one does not get out of anthropomorphism' (Canguilhem [1965] 2008, 8).[3] For scholars like Canguilhem and Bateson, this is a crucial recognition because it allows us to think the origin of the human as an ecological expression of life writ large, what Bateson ([1972] 2000, 1) calls an ecological 'conversation'. By this he does not mean thinking or communicating about ecological processes, but thinking and communicating as an ecological phenomenon.

## Life *in* the Anthropocene as general systematicity *of* life

I return to the question posed at the start of this chapter: what secures intervention as human authored if the genesis of the human is precisely what is at stake in determining the nature of life in the Anthropocene? So far, I have tried to problematise the notions of intervention and agency by pointing out that the self-evidence of the 'human' as the architect of nature or geology is what needs reexamination. But while I am trying to complicate the integrity of human identity as an autonomous entity, my aim is not to argue for a simple displacement or decentring of the human, for such a move still presumes a separation between the subject of human experience and the general ecology of life. Rather, I am persuaded by a much more involved sense of entanglement, following social theorists such as Durkheim, Luhmann and Bateson, that thoroughly confounds the ontology of the 'and' that is operative in the puzzle of relationality. If relationality is not an 'in-between' pre-existing entities given that the production of their difference, as Luhmann says, is what 'provides the system with systematicity' (Luhmann 1995a, 18), what then emerges is a radically reworked notion of causal agency, one that refigures the identity of the human, including all its cultural sophistication, as the world's self-encounter.

---

3   Like Bateson, Canguilhem makes a fascinating point that challenges knowledge or inquiry as a human achievement. He remarks in a wonderful passage: 'Doubtless, the animal cannot resolve all the problems we present to it, but this is because these problems are ours and not its own. Could man make a nest better than a bird, a web better than a spider? And if we look closely, does human thought manifest in its inventions an independence from the summons of need and the pressures of the milieu that would legitimate man's pity-tinged irony toward *infra*human living beings?' ([1965] 2008, xviii).

14

The point is not to deny or collapse the distinctions or hierarchies that are inherent to conceptions of the relationship between the human and whatever is defined as non-human. Rather, the more provocative suggestion offered by systems theorists like Luhmann and Bateson is that the perceived errors of human activity are aspects of life's determination to organise itself, understand itself, to be present to itself even in its misadventures. If life is its self-making, self-referential process, such a general ecology of movement *cannot* exclude its perceived negations, aberrations or mistakes. The broader significance of this point is crucial. For Luhmann, that every distinction implicates itself suggests that ethical discourse – expressing the rightness or wrongness of a given problem – is fundamentally compromised because it is premised on the 'interdiction of self-exemption' (Luhmann in Moeller 2006, 112). In other words, assuming a moral position does not acknowledge that that position is entangled with the very problem it deems unethical. For Luhmann, the force of this claim lies in the recognition that 'anyone who develops theories about "the" self develops theories about "his" self'. (Luhmann 1995a, 481). Here, if we read Luhmann's notion of general systematicity alongside Bateson's examples of the self-referential capacities of life, the sense of implication that is being evoked offers a more generous view of the nature of life as one that recollects itself in all its enduring diversities, tensions and apertures.

In this vein, the important insight that can be gleaned from Luhmann's account of the paradox of self-reference is precisely that a given intervention cannot be 'pure' or free of contamination from the purported mistake. They are one and the same. Yet, if we are persuaded by this reading, it becomes clear that what is at issue in the Anthropocene is the fundamental assumption that the intervening force and the object of error are separate 'events'. Without the recognition that the intervention and the error arise at the same moment, the notion of the Anthropocene strives and struggles to reject human exceptionalism while preserving the human species as the autonomous agent of responsibility and change. In other words, the human is regarded in the Anthropocene as at once the inquiring, intervening subject and the erroneous object whose environmentally damaging behaviour needs a corrective. Ironically, however, this corrective is fundamentally anthropocentric since the intervening subject is the object deserving of blame.

Crucially, here, what we need to confront honestly is the assumption that we can 'get out' or be rid of anthropocentrism as if it is not the reference point for which the non-human even makes sense. In short, any account of a human duty of care toward the non-human invariably secures and animates anthropomorphism/anthropocentrism.

By proposing that life is a self-referential system whose origin does not locate human agency but agency as systematicity, what emerges in Luhmann and Bateson's work is a very different way to engage the problem of the Anthropos, and along with it questions of intervention and activism. Stated differently, to render agency and intervention as expressions of life's self-encounter is not an attempt to diminish or to excise human agency. If life is self-referential, how justifications for ethical and political interventions in the Anthropocene are managed cannot be grounded in the reference point of the human against which other non-human species or the terrestrial environment is broached. What is needed is a new understanding of the difference between the culprit and the corrective that attests to their entanglement as expressions of *one* phenomenon: Life. Here, the question of what constitutes the mark of error or wrongdoing emerges as a general problem of location. As Luhmann says, 'every accident, every impulse, every error is productive ... Without "noise", no system' (Luhmann 1995a, 116). This is a radical complication of the classical view of cause and effect, or problem and solution, as a temporal distinction between before and after, as well as a spatial distinction between inside and outside.

Significantly, if life on Earth is self-referential, the problem of locating the human as the independent source of intervention (or the independent source of blame) is no longer tenable. Vicki Kirby's cogent interpretation in *Quantum anthropologies* (2011) of the 'place' of the human in an enlarged ecological understanding of the world speaks volumes here. If, Kirby argues, 'the Earth's grammar is necessarily internal, a shifting algorithm', it follows that

'the human' would not be bound and restricted by some special lack of access to that same generative unfolding. 'The human' would certainly be a unique determination, yet 'one' whose cacophonous reverberations would speak of earthly concerns. (Kirby 2011, 39)

Following Kirby, we might argue that what defines culpability or responsibility emerges in its inconsistencies and competing justifications as the moral or political quandary that morphs into different interventions and not as a compass that orients 'us' from a position outside the landscape of environmental devastation.

To be sure, this is not an invitation to relinquish our sense of personal agency because, after all, our intentions and interventions seem to be determined by the ongoing organisation of the world. Rather, it is precisely because, echoing Karen Barad, we are the world we seek to understand that our participation matters (Barad 2007, 396). 'The point', as Barad underlines, 'is not simply to put the observer or knower back *in* the world (as if the world were a container and we needed merely to acknowledge our situatedness in it)' (Barad 2007, 91, italics in original). Rather, she convincingly argues, '[w]e do not obtain knowledge by standing outside of the world; we know because "we" are *of* the world' (Barad 2007, 185, italics in original). This crucial acknowledgement that responsibility or agency is not secured 'in' a self-contained entity or species category will challenge the way we continue to address questions of blame, ethical accountability, and political intervention, questions that implicate us. In other words, 'the human' of the Anthropocene emerges, inevitably, as at once the agent of intervention yet also the object of error because the human is life lived as an ecological/geological entanglement.

## Works cited

Alberts P (2011). Responsibility towards life in the early Anthropocene. *Angelaki: Journal of the Theoretical Humanities* 16(4): 5–17.

Barad K (2007). *Meeting the universe halfway: quantum physics and the entanglement of matter and meaning.* Durham, NC & London: Duke University Press.

Bateson G. ([1972] 2000). *Steps to an ecology of mind: collected essays in anthropology, psychiatry, evolution, and epistemology.* Chicago, IL: University of Chicago Press.

Bateson G ([1979] 2002). *Mind and nature: a necessary unity.* London: Wildwood House.

Canguilhem G ([1965] 2008). *Knowledge of life.* P Marrati & T Meyers (Eds). S Geroulanos & D Ginsburg (Trans). New York: Fordham University Press.

Chakrabarty D (2009). The climate of history: four theses. *Critical Inquiry* 35(2): 197–222.

Chiew F (2014). Posthuman ethics with Cary Wolfe and Karen Barad: animal compassion as trans-species entanglement. *Theory, Culture & Society* 31(4): 51–69.

Crutzen P (2002). Geology of mankind. *Nature* 415: 23.

Dibley B (2012). 'The shape of things to come': seven theses on the Anthropocene and attachment. *Australian Humanities Review* 52: 139–53.

Durkheim É ([1895] 1982). *The rules of sociological method.* WD Halls (Trans). S Lukes (Ed). New York: The Free Press.

Durkheim É ([1914] 1973). The dualism of human nature and its social conditions. In R Bellah (Ed). *On morality and society: selected writings* (pp149–66). Chicago, IL: University of Chicago Press.

Kirby V (2011). *Quantum anthropologies: life at large.* Durham, NC & London: Duke University Press.

Latour B (2010). A plea for earthly sciences. In J Burnett, S Jeffers & G Thomas (Eds). *New social connections: sociology's subjects and objects* (pp72–84). Basingstoke, UK & New York: Palgrave Macmillan.

Luhmann N (1989). *Ecological communication.* Cambridge, UK: Polity Press.

Luhmann N (1993). Deconstruction as second-order observing. *New Literary History* 24(4): 763–82.

Luhmann N (1995a). *Social systems.* J Bednarz Jr & D Baecker (Trans). Stanford, CA: Stanford University Press.

Luhmann N (1995b). The paradoxy of observing systems. *Cultural Critique* 31: 37–55.

Luhmann N (1995c). Theory of a different order: a conversation with Katherine Hayles and Niklas Luhmann. *Cultural Critique* 31: 7–36.

Moeller H (2006). *Luhmann explained: from souls to systems.* Peru, IL: Open Court Publishing.

Rasch W (Ed) (2002). *Theories of distinction: redescribing the descriptions of modernity.* Stanford, CA: Stanford University Press.

Sayre NF (2012). The politics of the anthropogenic. *Annual Review of Anthropology* 41: 57–70.

Zalasiewicz J, Williams M, Steffen W & Crutzen P (2010). The new world of the Anthropocene. *American Chemical Society* 44(7): 2228–31.

Zalasiewicz J, Williams M, Haywood A & Ellis M (2011). The Anthropocene: a new epoch of geological time? *Philosophical Transactions of the Royal Society A: Mathematical, Physical and Engineering Sciences* 369(1938): 835–41.

# 2

# Anthropocene: the enigma of 'the geomorphic fold'

*Ben Dibley*

## The narcissistic subject

After the humiliating blows to the human subject delivered, as the cliché goes, by Copernicus, Darwin, and Freud, the notion of the Anthropocene would appear as some strange reversal, positing the Anthropos at the centre of planetary time and the Earth system's evolution. By some weird twist, it would seem 'man' is back – the human subject, returned to its privileged position, reassuringly ensconced at the hub of planetary disorder. The notion would seem to confirm at least the possibility of human control and dominance. This much is clear for literary critic, Tom Cohen, as he sardonically puts it: 'The term "anthropocene" ... [is] the epitome of anthropomorphism itself – irradiating with a secret pride invoking comments on our god-like powers and ownership of "the planet" ' (Cohen 2012, 240). The formulation in this chapter's title, 'the geomorphic fold', is borrowed from Cohen (2010).

The notion of the Anthropocene, it is true, already has a complicated relationship with the humanities and social sciences; one I am

Dibley B (2015). Anthropocene: the enigma of 'the geomorphic fold'. In Human Animal Research Network Editorial Collective (Eds). *Animals in the Anthropocene: critical perspectives on non-human futures*. Sydney: Sydney University Press.

sure, that will only become more so, and one that certainly cannot be reduced to the throwaway lines of critics (Dibley 2012a). Yet, in taking issue with the apparent re-centring evoked by the positing of the planetary agency of the human species, Cohen's (2012) is, no doubt, a familiar, and widely voiced objection. Such evaluations, which contend the Anthropocene is but the ominous height of anthropocentricism, I think, deserve a note of caution. This is not because they are misplaced in railing against the hubris of the 'God species' or the 'good Anthropocene', to cite recent journalistic formulations that would celebrate the possibility of successful human regulation of the Earth's systems and their management in practices of genetic modification and geo-engineering (Lynas 2011).[1] It is because the figure of the Anthropos nested in the Anthropocene is perhaps not quite the figure that humanities and social science scholars are accustomed to and might take as their target. And it is certainly not the sovereign subject in which nature is but its servant, resource and receptacle. It is a rather different (political) animal, at least in its scientific formulations.

Far from suturing a wounded human narcissism, this Anthropos – the Anthropos of the Anthropocene – is a paradoxical, enigmatic figure: one that we would do well not to take at face value. For as much as it might be a troubling re-inscription of human exceptionalism, it is one that simultaneously troubles any claim contending that it embodies this re-centring.

---

1   In a debate in the *New York Times*, 23 May 2011, commentators including ecologist Earle Ellis and libertarian journalist Ronald Bailey, contended the 'bad Anthropocene' ushered in by the Industrial Revolution might be overtaken by the 'good Anthropocene' as humanity, in becoming conscious of its own geological agency, takes active measures to control planetary systems through technical intervention (also see Revkin 2011). Futurist and transhumanist, Anders Sandberg, has echoed this notion of the 'good Anthropocene', contending we live in a charmed geological interval: 'The anthropocene is a magical era. It is up to us to design it so that it is a good world to live in for everyone and everything' (2012, np).

## The ends of the human

With the emergence of the human species as a geological force, it would be fair to say that the Anthropocene ramps up critical questions of human exceptionalism 'to a new level of intensity' (Alberts 2011, 9). A number of writers have remarked on the irony that, at the moment the Anthropocene dawns as a geological hypothesis positing the human species as a geo-agent forcing planetary limits, the leading edge of humanities and social science scholarship is concerned with moves to finally debunk human exceptionalism (Colebrook 2012a; Morton 2012; Latour 2013). When, through meticulous empirical research, scientists discover the human species is a geological force, explorations of the posthuman, of the more-than-human, radically reject the possibility of human uniqueness – that it is the bearer of any defining attributes, that we can know what the human is. Rather, on these accounts, the human becomes but a continuation of life, of matter. Human and non-human entities are 'not separate kinds of being' to be ordered hierarchically, rather they 'are all just different collections of the same stuff – bacteria, heavy metals, atoms, matter-energy' and so on (Bennett 2010, 122). They are, as Jane Bennett argues, entities sharing in a 'common materiality' (2010, 122). Concisely capturing the irony, Claire Colebrook, puts it thus: 'precisely when man ought to be a formidable presence, precisely when we should be confronting the fact that the human species is exceptional in its distinguishing power, we affirm that there is one single, interconnected, life-affirming ecological totality' (2012b, np). It would seem that there is afoot 'a strange double shift' (Colebrook 2012b, np).

During the last 200 years or so, just as the sediment of human action was forming on the Earth's surface, 'man', in 'his' humanist, and inclusive/occlusive guise, was deemed to have no essence other 'than the form [of] life he gave to himself' (Colebrook 2012b, np). No essence other than 'his' capacity or potentiality for self-conscious change that as Marx ([1844] 1959) would have it, distinguished the human animal from other animals as a 'species-for-itself' – as a 'species-being.'[2] Eman-

---

2    For a related discussion of the notion of species-being and the Anthropocene see Dibley (2012b). This provides a reading of how Marx's notion is mobilised in

cipated from any determining essence, there would seem to be a weird twofold alternation. On the one hand we face the figural 'death of Man' in which 'the human subject' is epistemologically extinguished: 'it is declared', as Colebrook sums it:

> that there really is no such thing as man, that the notion of human exceptionalism was a lie and that in truth there is one life in which all the features that had once marked the human – knowledge, emotion, linguistic capacity, altruism, mind and community – are in fact present in all life' (2012b np).

The human is disclosed to be a mistaken identity; the human being is, in Dominic Pettman's pithy pun, but a 'specious being' (Pettman 2011, 199).

On the other hand we face the literal 'death of Man', in which the human species is ecologically extinguished. While the Anthropocene has yet to run its course, it is declared that this epoch 'will be a standout event in the 4.5 billion year history of the Earth' (Mark Williams, cited in Falcon-Lang 2011, np). From the perspective of this prospective geology, the human age will be but a distinct stratum, legible 'in the Earth's layers from a post-human future'. As geologists Jan Zalasiewicz and colleagues sum it, the human is thus disclosed to be 'a stratigraphic interval' (Zalasiewicz et al. 2011, 6). The human here is a geological event, mineralised into the fossil record. And so, we are delivered to the strange paradox: that, while we cannot know the truth of the human being, it will be clear where the human has once been.

Among other things, it seems to me that the Anthropocene is an invitation to a difficult dance if we are to negotiate these ends of the human. For there would seem to be the need to walk a peculiar line between the recognition of one – that is, of the rise of a distributed humanity that with its prosthetics and animals, now composes a geological agent; while taking responsibility for the (legacy of the) other – that is, of the unintended consequences of the humanist subject's

---

discussions of climate change by Chakrabarty (2009), and Žižek (2010) among others.

emancipatory quest in whose wake ecological calamity has unfolded. Tentative steps here would seem to involve embracing the irony of this strange paradox that simultaneously demands, recognising that we know not what the human is, and, yet acknowledging that what is definitive about the human, will be inscribed in the stratigraphic record (Zalasiewicz 2008), literally written in stone.

## The enigmatic fold

In relation to the first half of this paradox Timothy Morton has drawn out the affinities between the scientific exposition of the Anthropocene and recent developments in continental philosophy, particularly, in his work on 'hyperobjects' (Morton 2010; 2012; 2013). Here he designates 'the gigantic nonhuman beings' – radioactive materials, global warming, ocean acidification, and so on that 'are massively distributed in time and space relative to humans', and, as such, as he continues, are 'the very script of the layers in Earth's crust that opens the Anthropocene' (Morton 2012, 232). In addition to the capacity for forcing planetary boundaries into novel states, one thing hyperobjects do is eliminate the ground once and for all of the pretence of the human as sovereign subject, as a privileged site of agency. This is not to deny that the species is a geological force. Rather it is to recognise other 'nonhuman actors as genuine "actors", without', as Levi Bryant writes, 'reducing them to props in dramas of human interest and without portraying ourselves as gracious sovereigns that wave our hands in acts of dispensation, deigning to concern ourselves with animals, rocks, planets, elements, etc., such that we "rescue" them from their reduction to our consumption' (2010, np). If anything is true, it is that hyperobjects are beyond such rescue. They now have a life of their own – one in the case of nuclear waste or climate change that will stretch for tens of thousands of years beyond the present. This is but a blink of the eye for the Earth and eons for the humans. As 'our lasting legacy' Morton contends 'the ecological thought must think the future of these objects' (2010, 130). In the face of the immense scale of these biogeochemical assemblages it is a thought that can only overturn 'any theological pretension to treat humans as the Lord, center, or master of being' (Bryant 2010, np). It is also a thought that requires the recognition of responsibility in the face of these hyperobjects. Having birthed such monstrous entities they now return demanding care from their maker lest they destroy their progenitor (Latour 2012).

Yet, and to turn to the other side of the paradox, while such accounts of distributed agency supply important leverage on the ways in which the non-human is inextricable from the human and which the unfolding of the Anthropocene has made increasingly discernible, the advent of the human species as a geological event does raise the question: what is definitive to the human and to the Earth that would see the species emerge as this force? If, as the Anthropocene hypothesis postulates the ' "earth" has evolved to produce a species capable of altering the earth's own limits' (Colebrook 2012a, 203) – a twofold question would seem to follow: firstly, if the Anthropocene witnesses the unleashing of hyperobjects that are in the process of altering planetary boundaries, it surely is worth asking, with the environmental geographer, Nigel Clark: *what kind of planet is this that births a creature capable of doing such things?* (2012, 261, italics in original). And, subsequently: what might the definitive characteristics of this creature be, who has this capacity to force the Earth's parametric boundaries? If humanism and its successor posthumanism come to posit the human as that without a determining essence – the techno-species without qualities (Agamben 2000), the species without an ecological niche (Virno 2004) – the unfolding of the Anthropocene might well be pause for (the ecological) thought.

## Geostory and the fire species

Before taking up such questions directly it is useful to turn to Bruno Latour. In his recent Gifford Lectures, Latour (2013) offers his fullest exposition to date of the Anthropocene. Given that the geologists' account of the Anthropocene leaves behind just about everything Latour (1993) has had to say about the 'Modern Constitution' – that arrangement that maintained 'the great divide' between nature and society – it is perhaps unsurprising that he finds in the Anthropocene an alternative to 'the modern', an alternative to theories of modernity. Central to his recent exposition is the notion of Geostory. In contrast to the narrative of the moderns, of 'human subjects', that are in a struggle to free themselves from all that would limit the potentiality of their species-being, Geostory is the narrative of the increasing entanglement of the human and the non-human, the unintended consequence of this emancipatory struggle, that the geologists now trace in processes that come to compose the Anthropocene.

In Latour's hands, however, Geostory is not only a counter-narrative to modernity as Earth system scientists make legible the increasingly dense loops entangling the human and the non-human in their accounts of climate change, ocean acidification, mass extinction, and so on. It is also a speculative opportunity in political theory, albeit a bleak one. Inasmuch as the moderns in their emancipatory quest fail to acknowledge the unfolding Anthropocene, it would appear that it is an entity without a public, an entity without a demos. As if they had another planet to which to escape the moderns have refused their attachment to this one, to this expansive biogeochemical assemblage that is the biosphere within which they are enveloped. In Latour's revamping of James Lovelock's (in)famous formulation the moderns refuse to face Gaia.

Objecting to those that object to the seeming anthropocentrism of the notion of the Anthropocene, Latour retorts: 'Not only should the Earth be the centre of our exclusive attention, but we should also feel responsible for what happens' (2013, 56). It is this 'we' – the earthlings, or in his most recent coinage, an emergent 'Earthbound', that he darkly projects in a Schmittian formulation, will come to be in 'a state of war' with the moderns as they come to the defence of 'the Earth', or, in Latour's new political theology, of Gaia (2013, 98–122).

As important and as sobering as it is, Geostory is not the whole story. Contending cosmograms pitted against one another is a necessary, but not exhaustive, diagram of the Anthropocene. This is so I think along two dimensions. Firstly, it is worth stressing that the Anthropocene inasmuch as it is an interval of geological time is not just about the humans and their relations with their non-human biogeochemical others and whatever it is that might succeed them in a posthuman polity to come. While the scientists' painstaking work comes to reveal, in Latour's witticism, an *anthropomorphism* on steroids' in which the traces of human action can be seen literally everywhere (2011, 3, italics in original); we need to take heed that in its scientific exposition the Anthropocene is as much about the decentring of the human as it is about the escalating geological agency of the human species, it is in some important sense *not about the human at all*. That is, if that figure is conceived as that without a determining essence, other than its potentiality for self-change – the humanist bequest to posthumanism. This is so, since, in strictly geological terms, if the Anthropocene was recognised as a formal stratigraphic interval, this figure is one of little import. As Bronislaw Szerszynski observes:

> it is important to realise that the truth of the Anthropocene is less about what humanity is doing, than the traces that humanity will leave behind. In terms of environmental ethics, one might say that geology is brutally consequentialist – it does not matter what one does, or why one did it, just what consequences it will leave behind (2012, 169).

Further it is important to note that explorations of such traces and the novel geology that they signal present geologists with further incitements 'to explore analogies, continuities and discontinuities across a range of [geological] epochs, most of which are *unequivocally inhuman*' (N Clark 2012, 260, my italics). An observation that has led Clark to argue that rather than seeing the Anthropocene as 'another incentive to decree the "end of nature" ':

> A more generous response of the humanities and social sciences to the scientific acknowledgement of human geologic agency . . . would be to join natural sciences in confronting the full range of geologic forces, without which 'our' agency would be an abstract and orphan presence in the universe. And this implies engaging with physical forces 'in themselves', and not simply 'for us'. (N Clark 2012, 260)

Second, then, I wonder if Geostory may have a prehistory or rather, it may have connections to a past longer that the 400 years or so of modernity to which it would be an alternative narrative, or, to the 8000 years of agriculture in whose beginnings some would seed the Anthropocene (Ruddiman 2003). To make this point I return to the question of whether there is anything definitive to the Anthropos of the Anthropocene, to this species that is now a geo-force among others. In advancing his own project of 'speculative geo-physics', Nigel Clark's (2012) tentative answer to these questions is in terms of that species' affinity with fire on a planet rich with oxygen, fuel and ignition sources. While humans have advanced a technics of the flame, from cooking to the combustion engine, to harness the energy of bio-materials, Clark draws on environmental historian, Stephen Pyne's (1997) more radical contention that 'fire use may be' the human species 'biological and geological niche' (N Clark 2012, 269).[3] Given that 'humans are the only life form in the planet's history to control fire', the human species might be the 'fire species'; we are, as Pyne puts it, 'fire creatures on a uniquely fire planet' (1997, 3 cited in N Clark 2012, 269). Glossing Pyne's argument, Clark writes: 'Though human-modulated combustion may be unique, Pyne views this as an extrapolation of terrestrial tendencies rather than as an anomaly: human fire use "accelerated, . . . animated, leveraged" what was already present' (Pyne 1997, 302–3; N Clark 2012, 269). 'Or to put it another way', Clark continues, 'humans [have] appropriated and advanced a technics that was the planet's own' (2012, 269). On this account, Geostory might be viewed as that concerned with the human species' long augmentation of the planet's own tendencies for combustion, of which a fossil-fuelled modernity is arguably the most enhancing, and certainly the most endangering to that species, as it is to many others.

## The detrivorous turn

However, this is not to give a false unity to the agency of the fire species as the universal agent of Geostory. While our hominid affinity with the flame might well be the defining quantity of the species and its biological and geological niche, the particular form of combustion preferred

---

3   In this connection there is an emerging geology literature that establishes the human species' 'mastery of fire' about 1.8 million years ago as the threshold of 'the Palaeoanthropocene'. See Andrew 2013.

by the moderns – 'the "digging up the dead," by bringing to the surface the fossilised, organic matter of once-living things and burning it' –
marks a mutation in the co-evolution of the human and Earth systems
(Griffiths 2010, np). William Catton (1980, 2009) proposes the term,
*Homo colossus*, for contemporary humans that like earthworms and
slugs, sustain themselves by breaking down an exhaustible accumulation of dead plant and animal material, releasing it again as energy into
the ecosystem. And like other detrivore communities that also rely on
finite reserves of dead organic matter for their livelihood, *Homo colossus*, for Catton, is caught in a 'bloom and crash' cycle – 'flourish[ing]
and collaps[ing] because they lack the life-sustaining biogeochemical
circularity of other kinds of ecosystems' (Catton 1980, 172). While no
doubt raising the prospect of a posthuman account of the contemporary 'double crisis' – of ecology and capital accumulation, the fire
species' detrivorous turn does present as something of an evolutionary
dead end.[4]


## Learning how to die in the Anthropocene

It is a perplexing paradox that knowledge of the Anthropocene and the
attendant risks of tipping its systems into states catastrophic for humans
and their sustaining biosphere has been accompanied by 'repeated government and social failures to do anything other than act in ways which
intensify the danger' (T Clark 2012, v). This has led Timothy Clark to
wryly remark: 'The actions of humanity considered en masse do not
just appear irrational, but in some ways so staggeringly stupid as to
constitute – among other things – a new philosophical problem' (vi).
While Clark's comments were made in an introduction to a collection
of essays on deconstruction and climate change, given over to the ex-

---

4   It is worth repeating Chakrabarty's counterfactual that should socialism have
won out, the Earth's system, particularly its climate, would be in a worse state than
it currently is, since the emancipation of the world's poor would have required
burning more fossil fuels than used by the current regime. Nevertheless, for an
important provocation on the prospects of 'geo-communism' see Arun Saldanha
2013.

plication of this 'suicidal slumber', it is perhaps Roy Scranton's recent opinion piece in the *New York Times*, 'Learning how to die in the Anthropocene' that most starkly poses that question:

> The biggest problem we face is a philosophical one: understanding that this civilisation is already dead. The sooner we confront this problem, and the sooner we realise there's nothing we can do to save ourselves, the sooner we can get down to the hard work of adapting, with mortal humility, to our new reality. (2013, np)

Resonating with the 'enlightened catastrophism' of Jean-Pierre Dupuy and Slavoj Žižek, in which the inevitability of the catastrophe is acknowledged so to open the possibility of new courses of action in the present, Scranton's morbid pedagogics proposes that if we are to forestall our own end, we must recognise we are already dead. Perhaps our only choice is, as Szerszynski plainly puts it, 'which kind of extinction of the human we are prepared to let happen: an ontic or an ontological one' (2010, 17).

## Works cited

Agamben G (2000). *The open: man and animal*. Stanford, CA: Stanford University Press.

Alberts P (2011). Responsibility towards life in the early Anthropocene. *Angelaki: Journal of the Theoretical Humanities* 16(4): 5–17.

Andrew G (2013). Fire and human evolution: the deep-time blueprints of the Anthropocene. *Anthropocene* 3: 89–92. Retrieved on 20 March 2014 from dx.doi.org/10.1016/j.ancene.2014.02.002.

Bennett J (2010). *Vibrant matter: a political ecology of things*. Durham, NC & London: Duke University Press.

Bryant L (2010). Inhuman ethics. *Larval Subject*, 20 January. Retrieved on 30 April 2015 from http://larvalsubjects.wordpress.com/2010/01/20/inhuman-ethics/.

Catton W (1980). *Overshoot: the ecological basis of revolutionary change*. Urbana, IL: University of Illinois Press.

Catton W (2009). *Bottleneck: humanity's impending impasse*. Bloomington, IN: Xlibris Corporation.

Chakrabarty D (2009). The climate of history: four theses. *Critical Inquiry* 39: 197–222.

Clark N (2012). Rock, life, fire: speculative geophysics and the Anthropocene. *The Oxford Literary Review* 34(2): 259–76.

Clark T (2012). Deconstruction in the Anthropocene. *Oxford Literary Review* 34(2): v–vi.

Cohen T (2010). The geomorphic fold: anapocalyptics, changing climes and 'late' deconstruction. *Oxford Literary Review* 32(1): 71–89.

Cohen T (2012). Polemos: 'I am at war with myself' or, Deconstruction™ in the Anthropocene? *Oxford Literary Review* 34(2): 239–57.

Colebrook C (2012a). Not symbiosis, not now: why anthropogenic change is not really human. *Oxford Literary Review* 34(2): 185–209.

Colebrook C (2012b). Introduction: extinction. Framing the end of the species. In C Colebrook (Ed). *Extinction* (np). Open Humanities Press. Retrieved on 30 April 2015 from http://www.livingbooksaboutlife.org/books/Extinction/Introduction.

Dibley B (2012a). 'The shape of things to come:' seven theses on the Anthropocene and attachment. *Australian Humanities Review* 52: 139–58.

Dibley B (2012b). 'Nature is us': the Anthropocene and species-being. *Transformations* 21 (Special Issue: Rethinking the seasons: new approaches to nature). Retrieved on 30 April 2015 from http://www.transformationsjournal.org/journal/issue_21/article_07.shtml.

Falcon-Lang H (2011). Anthropocene: have humans created a new geological age? *BBC News: Science and Environment*, 11 May. Retrieved on 30 April 2015 from http://www.bbc.co.uk/news/science-environment-13335683.

Griffiths T (2010). A humanist on thin ice. *Griffith Review* 29 (np).

Latour B (1993). *We have never been modern*. C Porter (Trans). Cambridge, MA: Harvard University Press.

Latour B (2011). Waiting for Gaia: composing the common world through arts and politics. A lecture at the French Institute, London, November 2011. Retrieved on 30 April 2015 from http://www.bruno-latour.fr/sites/default/files/124-GAIA-LONDON-SPEAP_0.pdf.

Latour B (2012). Love your monsters: why we must care for our technologies as we do our children. *The Breakthrough*, Winter. Retrieved on 21 April 2015 from http://thebreakthrough.org/index.php/journal/past-issues/issue-2/love-your-monsters

Latour B (2013). Facing Gaia: six lectures on the political theology of nature. The Gifford Lectures of Natural Religion Edinburgh, 18–28 February 2013.

Lynas M (2011) *The God species: how the planet can survive the age of humans*. London: Fourth Estate.

Marx K (1959). *Economic and philosophic manuscripts of 1844.* M Mulligan
(Trans). Moscow: Progress Publishers. Retrieved on 30 April 2015 from
https://www.marxists.org/archive/marx/works/1844/manuscripts/
preface.htm.

Morton T (2010). *The ecological thought.* Cambridge, MA: Harvard University
Press

Morton T (2012). Ecology without the present. *Oxford Literary Review* 34(2):
229–38.

Morton T (2013). *Hyperobjects: philosophy and ecology after the end of the world,*
Minneapolis, MN: University of Minnesota Press.

Pettman D (2011). *Human error: species-being and media machines.* Minneapolis,
MN: University of Minnesota Press.

Pyne S (1997). *World fire: the culture of fire on earth.* Washington, DC: University
of Washington Press.

Revkin A (2011). Embracing the Anthropocene. *New York Times,* 20 May.
Retrieved on 30 April 2015 from http://dotearth.blogs.nytimes.com/2011/05/
20/embracing-the-anthropocene/.

Ruddiman W (2003). The anthropogenic greenhouse era began thousands of years
ago. *Climate Change* 61: 261–93.

Saldanha A (2013). Some principles of geocommunism. *Geocritique,* 23 July.
Retrieved on 30 April 2015 from http://www.geocritique.org/
arun-saldanha-some-principles-of-geocommunism/.

Sandberg A (2012). Commercializing the robot ecosystem in the Anthropocene.
Future of Humanity Institute, University of Oxford. Notes for a talk given at
the Robotics Innovation Challenge, 9 February. Retrieved on 30 April 2015
from http://tiny.cc/y4b5xx.

Scranton R (2013). Learning how to die in the Anthropocene. *New York Times,* 10
November. Retrieved on 30 April 2015 from
http://www.opinionator.blogs.nytimes.com/2013/11/10/
learning-how-to-die-in-the-anthropocene/.

Szerszynski B (2010). Reading and writing the weather climate technics and the
moment of responsibility. *Theory, Culture & Society* 27(2–3): 9–30.

Szerszynski B (2012). The end of the end of nature: the Anthropocene and the fate
of the human. *Oxford Literary Review* 34(2): 165–84.

Virno P (2004). *A grammar of the multitude.* New York: Semiotext(e).

Zalasiewicz J (2008). *The earth after us: what legacy will humans leave in the rocks?*
Oxford, UK & New York: Oxford University Press.

Zalasiewicz J, Williams M, Smith A et al. (2011). Are we now living in the
Anthropocene? *GSA Today* 18(2): 4–8.

Ziarek K (2011). The limits of life. *Angelaki: Journal of the Theoretical Humanities* 16(4): 19–30.

Žižek S (2010). *Living in the end times*. London & New York: Verso.

# 3

# Cycles of anthropocenic interdependencies on the island of Cyprus

*Agata Mrva-Montoya*

The Anthropocene emphasises the central role of humankind in the alteration of atmospheric, biospheric, geologic, hydrologic and other ecological processes to a degree that raise concerns about the future of Earth environment and human civilisation. As the major geological force affecting the Earth systems, humankind has the responsibility for shaping the future of life on Earth by becoming, in the words of Will Steffen, Paul Crutzen and John McNeill, the 'Stewards of the Earth System', but, as they ask, 'Can humanity face this challenge?' (Steffen et al. 2007, 619). The degree of human impact is unprecedented in the history of human civilisation. In the last 150 years or so humans became geological agents affecting Earth's environment on a global scale (Zalasiewicz et al. 2010). But earlier societies were also transforming local ecosystems, sometimes irrevocably (Budiansky 1994). The view of pre-modern societies as 'biological agents influencing the environment' (Solli et al. 2011, 50) challenges the notion of their living in worshipful respect of nature and complicates the debate about the proposed timing of the Anthropocene.

Mrva-Montoya A (2015). Cycles of anthropocenic interdependencies on the island of Cyprus. In Human Animal Research Network Editorial Collective (Eds). *Animals in the Anthropocene: critical perspectives on non-human futures*. Sydney: Sydney University Press.

The onset of the Anthropocene is usually linked with the beginning of the Industrial Revolution c. 1800 (Steffen et al. 2007, 614),[1] but the results of archaeological research attest to humans making large-scale transformations of the Earth's ecosystem long before the 19th century. Some archaeologists point to the exodus of modern humans from Africa about 60,000 years ago and the beginning of large-scale hunting, clearing forests and burning of vegetation as the forerunner of 'human domination of the earth' (Balter 2013, 261). Human hunting has been linked with at least some of the megafauna extinctions, which has otherwise been attributed to the consequences of climate change. Others link the beginning of agriculture about 11,500 years ago at the beginning of the Holocene with dramatic impacts on the Earth's ecosystem (Ruddiman 2003, 2013; Balter 2013, 261; Rick et al. 2013; Smith & Zeder 2013).

With a long-term view of the interdependencies of human societies and the rest of nature, archaeology has the capacity to deepen our understanding of human intervention into the natural world and to offer a critical perspective on the capacity of humans to 'fix' or mitigate anthropocenic damage. Cyprus, a small island in the eastern Mediterranean, provides a unique case study of human impact on its fauna through the centuries and an argument in support of the 'early Anthropocene' theory proposed by William Ruddiman (2003, 2013) and others (Smith & Zeder 2013; Rick et al. 2013).

After the demise of Pleistocene megafauna, Cyprus was devoid of any larger mammals and the earliest human settlement by the end of the ninth millennium BC had a profound effect on the distribution of terrestrial species through either deliberate or, in the case of rodents, accidental introduction. Cypriots transformed the world of terrestrial non-volant animals on the island and have been shaping its ecosystem for more than 10,000 years. Over time some species established feral populations, some died out, new species were introduced and the cycle of introduction, feralisation, naturalisation, control and extinction continues to the present. The socio-economic, cultural, political and technological developments continue to deeply affect the relationships be-

---

1   Unless specified as BP (before present) – usually calibrated (cal) – or BC (before Christ), the dates refer to the current era.

tween human and animal world, and all too often animals bear the brunt of human-made disasters, financial and political crises, and sheer folly.

The future of animals on anthropocenic Cyprus remains contingent on the ongoing human engineering of nature, disguised as animal conservation and management. The fate of individual species is generally linked with the traditional classification of animal species into dichotomies of domestic/wild and native/alien. These divisions, however, have remained permeable since antiquity and are further complicated by the changing cultural, political and legal contexts. These seem to change faster than the societal values and individual behaviour of Cypriots, a trend symptomatic of the main challenge facing the world in the Anthropocene: is humanity capable of changing fast enough to mitigate further climate change and loss of biodiversity?

Using Cyprus as a case study of human management of species life, this chapter aims to show the complexity of anthropocenic interdependencies and a long-term pedigree of human intervention into the 'natural world'. The ongoing cycle of human engineering of the animal world in Cyprus carries implications for the current animal conservation and management issues on the island and elsewhere, and the wider concept of 'taking responsibility' for anthropocenic damage in general.

## Pleistocene colonisations and extinctions

Cyprus, one of the more isolated islands of the Mediterranean, rose from the sea floor during the Miocene period (Held 1989). According to the geological evidence there was never a land bridge connecting Cyprus to the mainland. Even during the Pleistocene glaciations, affecting the water levels in the Mediterranean, the island remained separated from the mainland by at least 30 km to 40 km of deep water (Simmons 1999, 27). The origin of the island, its environmental conditions and the restrictions imposed by its geographical location played a major role in influencing the distribution of terrestrial animal species before human colonisation. The sea acted as a filter, letting through only those animal species that could swim or that had accidentally completed a sea crossing effected, for example, by rafting on uprooted trees and rafts of vegetation torn up from the mainland during storms and hurricanes, or other 'chance combinations of favorable circumstances' (Masseti 2009a, 170). Not surpisingly, Pleistocene fauna of Cyprus has been described as 'oceanic in its paucity' (Boekschoten & Sondaar 1972, 333). The Pleistocene glaciations and climate change created fluctuations of the sea levels, which 'favoured *sweepstake* migrations, and the subsequent evolution of endemic faunas, by altering the distances between the mainland and the islands' (Masseti 2009a, 170, italics in original).

The founder populations of elephants and hippopotamuses must have arrived on Cyprus when the sea level was low during one of the major Middle Pleistocene glacial episodes,[2] possibly as a result of an accidental sea crossing with no opportunities of return (Sondaar 1991, 252). Fossil remains of both herbivores show signs of morphological change, a result of isolation and environmental adaptation, with the size being the most obvious one. Compared to the modern living African elephant with a shoulder height of 3 to 4 metres, the Cypriot pygmy elephant (*Elephas cypriotes*) was probably about 1 metre tall. In the case of the pygmy hippopotamus (*Phanourios minutus*), apart from a smaller size (about 1.5 metres long and 0.75 metres tall), other morphological changes visible in the skeleton indicate that the animal was adapted to walking rather than swimming (Reese 1993, 320; Simmons 1999, 29).

Apart from herds of the pygmy hippopotamus and a small number of the pygmy elephants, the Pleistocene fossil record also contains a genet (*Genetta plesictoides*), a mouse and a soricid (Held 1992, 112; Reese 1995). The presence of a genet (a viverrid feline-like carnivore with a spotted coat, a very long banded tail, a small head and large ears) is surprising as carnivores lack the floating capacity and herd-behaviour of herbivores, and are usually absent in insular endemic faunas of 'sweepstake' origin (Theodorou et al. 2007). The Cypriot white-toothed shrew (*Crocidurs suaveolens praecypria*) probably colonised the island during the Lower Pleistocene (1.54 million years ago), while the

---

2    The lowest level would have existed about 380,000 (Mindel Glaciations) and between 195,000 and 165,000 (Riss Glaciations) years ago (Swiny 1988, 7). Reese put the possible date of their arrival in the Later Pleistocene between 250,000 and 100,000 years ago (Reese 1992b, 50).

Cypriot mouse (*Mus cypriacus*) may have arrived 0.5 to 1.0 million years ago (Cucchi et al. 2006; Dubey et al. 2007, 3449). These rodents, and a few species of snakes and amphibians, are the only terrestrial animals that arrived in Cyprus without human interference and have survived to modern times.

The absence of larger carnivores and resource competitors contributed to the survival of hippopotamuses and elephants into the early Holocene and then they died out. The demise of megafauna was variously linked with uncontrolled population growth in a predator-free environment, resulting in overgrazing and subsequent starvation (Simmons 1999, 29–30), with climate and habitat changes at the end of the Pleistocene (Diamond 1992, 15), and finally with the impact of early human settlement on the island.

The impact of humans on the extinction of pygmy animals has been hotly debated since the discovery of hippopotamus remains at Akrotiri Aetokremnos and the interpretation remains, to a certain extent, inconclusive. The debate concentrates on the character of the deposition – whether it was natural or anthropogenic.[3] The animal remains recovered during excavations of the site in 1987, 1988 and 1990 yielded more than 200,000 bones of which more than 98 percent belonged to the pygmy hippopotamus, representing more than 500 individuals of different ages. Apart from hippopotamuses, bones of numerous birds, three elephants, and remains of pigs, genets, murids, tortoises, snakes and toads were found. The faunal remains were accompanied by more than 1000 chipped stone artefacts including blades, flakes, small thumbnail scrapers, a few ground-stone implements and numerous stone and shell beads. The series of radiocarbon dates acquired from the site point to a short-term occupation during the tenth millennium BC, lasting possibly a few hundred years and centred around the calibrated calendar age of 9825 BC (Simmons 1999, 153–69, 186–87; Simmons 2001, 4–5).

Simmons argues that a hunter-gatherer community of fewer than 50 people could have lived on the peninsula and possibly hunted to extinction the population of hippopotamus in the area, already weakened

---

3    See summary of implications and arguments in Simmons 1992, 353–54, tables 1 and 2.

by the climatic and environmental deterioration at the end of the Pleistocene era. Interestingly, the number of bird bones and shells increases while the amount of hippopotamus remains decreases in the upper layers, suggesting a need for other sources of meat when the herbivores became scarce (Sondaar & Van der Greer 2000; Simmons 2001, 12). Such shifts in consumption of animal species, and, more importantly, waves of extinctions of indigenous fauna and introductions of new taxa following human colonisation is known from other islands (Rick et al. 2013).

The extinction of the Pleistocene megafauna left Cyprus devoid of larger mammalian herbivores. This situation was soon changed with the first settlers bringing over numerous species of animals as a 'living larder' in a process not unlike the various 19th-century Australian acclimatisation societies that aimed to ' "introduce, acclimatize, domesticate and then liberate select animal, insect, and bird species from England" that would transform the colony in the home country's image and diversify the economic base of . . . agriculture' (Anderson 1998, 36). This was the start of the cycle of anthropogenic animal introductions that was responsible for a dramatic increase of biodiversity on the island, which has continued to the present day.

## Animal translocations

A boar (*Sus scrofa*) was one of the first animal species brought to the island during or even before the 12th millennium cal BP. It seems that boars were introduced as wild animals and were hunted. In the recently excavated early farming settlements dated to ~11,100 to 10,600 cal BP, remains of small domestic dogs (*Canis familiaris*) and cats (*Felis lybica*) were also found, apart from bones of wild boar. Dogs could have been used in hunting, while cats may have been used to protect crops from rodents. Both species may have contributed to the extinction of the genet (Vigne et al. 2009; 2011, S260; 2012, 8448).

Animal remains found at the early Pre-Pottery Neolithic B site of Parekklisha Shillourokambos from c. 10,4000/10,300 cal BP to c. 9000 BP provide a snapshot of more introductions and changing patterns in the use of animals. At first, the inhabitants relied on hunting of autochthonous wild boar. They also introduced an early domestic form of goat (*Capra aegagrus/hircus*), domestic cattle (*Bos primigenius/taurus*) and fox (*Vulpes vulpes*). While goats rapidly spread throughout the island and were hunted, the cattle were bred and remained of low importance in the meat supply throughout the tenth millennium BP (Vigne et al. 2011, S261, S263).

The over-reliance on boar-hunting led to demographic erosion of the wild population over time, but people continued to hunt goats and the Mesopotamian fallow deer (*Dama mesopotamica*), introduced from the southern Levant (Vigne et al. 2011, S262). Apart from wild deer, domestic sheep (*Ovis aries*) were introduced and bred for meat and milk. In contrast to goat, the result of morphological analyses and the slaughtering profiles of the animals show 'no evidence for feralization or sheep hunting at Shillourokambos' (Vigne et al. 2011, S264).

Apart from sheep, new lineages of domestic pig and cattle were introduced in Shillourokambos in the course of the tenth millennium BP, brought from other parts of the island or from the mainland. Over time people stopped hunting wild boar and goat, possibly due to reduction of wild populations, and relied instead on breeding for their meat supply. Domestic goats were used for milk production, while sheep provided meat and wool (Vigne et al. 2011, S263, S264).

The Mesopotamian fallow deer was never domesticated in Cyprus, even though it acquired an unprecedented economic importance on the island (Croft 2002). The importance of deer increased in the Late (Pottery) Neolithic (5500 to 4000 BC) and Chalcolithic (4000 to 2300 BC) sites where they made up to 75 percent of the animal remains. Throughout the Chalcolithic the importance of deer decreased in the sites of the Ktima Lowlands, but they still remained a significant component of the diet, and antlers were very important in the production of tools. By the Late Chalcolithic deer-hunting and game consumption declined and seem to have acquired a prestigious character in Kissonerga Mosphilia. The gradual decline in the importance of deer was associated with the degradation of the local environment and overhunting (Croft 1991, 199–200; Peltenburg 1998, 253; Marczewska 2005).

Equids were probably introduced to Cyprus in the early part of the Bronze Age (c. 2300 BC). As only small quantities of bone material are ever present in a settlement context and usually do not show evidence of butchery, it seems that the role of equids as either a beast of burden or a steed was far more important than as a dietary contribution. Moreover, horse and donkeys also rarely appear in a sanctuary context, indicating that they were not considered as a species suitable for a cult-related sacrifice, though they appeared in tombs dated to Early Cypriot–Middle Cypriot (2400 to 1650 BC) and Late Cypriot (1650 to 1050 BC) periods, and later in Cypro Geometric III–Cypro Archaic (900 to 475 BC) periods (Mrva-Montoya 2013). The presence of feral equids in ancient Cyprus is uncertain, but possible in the case of donkeys. Horses were expensive animals to keep and it is unlikely that they were purposefully released in order to form wild populations. They also have much higher food and water requirements than donkeys, thus precluding the naturalisation of the species in the arid climate of Cyprus.

Apart from active introduction of domesticated animals and game species, humans involuntarily introduced several small mammalian genera on Cyprus and other islands of the Mediterranean. Commensal species (such as *Mus*, *Rattus* or *Acomys*) and synanthropic ones (such as *Crocidura*, *Suncus* or *Apodemus*) most likely reached the islands by passive boat transport. The house mouse (*Mus musculus domesticus*) was introduced to Cyprus at least by 10,400 cal BP. The similarity between mice in Cyprus and on the mainland indicates 'a constant and intensive genetic flow from the mainland to Cyprus from the Neolithic up to today', as frequent as several times a year (Vigne et al. 2011, S264; 2012, 8448).

As can be seen from this overview, ancient Cypriots were instrumental in the formation of the mammalian world on the island (both wild and domestic) that, with a few exceptions (for example, camels were introduced at some point and the Mesopotamian fallow deer died out), persists to the present day. The early stages of human colonisation of Cyprus coincided with the Neolithic revolution in the Near East, which according to Ruddiman (2003, 2013) marked the beginning of longer-term anthropogenic changes and had a significant impact on greenhouse gas emissions, and according to Smith and Zeder (2013) marked the pivotal point in humans' ability to modify the ecosystem. The scale of human intervention into the animal world of Cyprus fits well with the early Anthropocene theory. In the context of animal world of Cyprus, the Anthropocene started with the arrival of humans at the beginning of the Holocene. People were responsible for colonising Cyprus with a variety of animals, both domestic and wild.


## Dynamic equilibrium

While the understanding of the introduction patterns and the transition from hunting to stock-keeping keeps changing as new sites are excavated and animal remains are analysed, it is clear that the distinction between wild and domestic animals in ancient Cyprus is problematic. In an ideal world, a classification system is complete, based on consistent principles and with mutually exclusive categories (Bowker & Star 1999, 10). It is based on science and remains constant. In the real world, and the world of Cypriot animals, the classification system is flexible, fluid and culturally constructed.

The introduction of animals in Cyprus took place concurrently with the early stages of animal domestication on the mainland. In fact the discovery of cats on the island, out of their natural area of distribution, pushed back the date of earliest human–feline association to the 11th millennium cal BP. Although humans were responsible for bringing all large terrestrial non-volant animals in the Neolithic, the exact nature of the human–animal relationship remains difficult to ascertain. Were these wild, 'managed', pre-domestic, domestic or feral animals? These categories are further complicated by interbreeding which most likely occurred as both bred and free-range animals belonged to the same species.

The ability to differentiate between domestic and wild animals in the archaeological record relies on the presence of biological modifications resulting from domestication, which varies in different species and is dependent on 'the intensity and nature of the relationship' (Vigne 2011, 173). While in the later periods the morphological differences are more discernible, the evidence is limited, especially for Iron Age Cyprus (1050 to 300 BC), reflecting the odds of archaeological preservation and the diverse methodologies of recovery and analysis of the bone material. Apart from morphological changes, zooarchaeologists are looking for the presence of animals in human burials, the frequency the skeletal parts and specific species in food refuse and the signs of selective killing-off patterns (age and sex) to distinguish between the different categories (Vigne 2011, 173; Rowley-Conwy et al. 2012). They are trying to impose a scientifically based and fixed classification system on incomplete records. These escape exact categorisation in the first place.

As Vigne and others have pointed out, domestication is a continuous process (Driscoll et al. 2009, 9972; Vigne 2011, 172). It can take various forms that:

> can be arranged on a gradient of eco-anthropological mutualistic relationships between animal and human societies . . . from anthropophily, to commensalism or control in the wild, the management of captive animals, expansive or intensive breeding, and finally to pets . . . As the process depends solely on the dynamic equilibrium between animals and humans, it is possible to achieve sustained stability at any level; with further progression or the retreat back to a less intensive relationship being possible. (Vigne 2011, 172–73)

Many of the animals introduced in Cyprus became naturalised, that is, 'they have formed self-supporting populations capable of perpetuating themselves' (Sax & Gaines 2008, 11491). Animals may have formed feral populations unaided, but this process may also have been encouraged by humans with animals being purposefully released to establish 'wild' game. That seems to have been the case with boar, deer, hare, foxes and goats. Animals may also have been bred in a free-ranging state. In modern times, this type of husbandry was practised until the 1960s in many of the Aegean islands where sheep, goats and even horses were let loose for most of the year and were caught when required, for food or for work (Masseti 1998, 12). In Cyprus, the population of feral donkeys in the Karpas Peninsula may go back to the 18th-century custom (or even earlier) of letting donkeys roam free to fend for themselves in winter when they were not needed for work and food was scarce. In summer some animals were caught and were tamed for the season's work, but a number of individuals must have turned feral and established a 'wild' population. Wild donkeys appear in textual evidence for the first time in the 13th century (Reese 1992a, 16).

Feral goats were reported as extinct in the 18th century (Reese 1992a), but they may have been present on the island as late as the 1930s (Croft 1991, 67). In the late 19th century the population of domestic goats was estimated to be the largest in the Mediterranean islands in relation to the area and population of Cyprus. Apparently there were about 250,000 goats, 150 goats per square mile (57.7 goats per square km) and almost two goats per inhabitant and had a devastating impact on the island's vegetation. The expansion of goat pastoralism and uncontrolled grazing was linked with the decay of agriculture and depopulation during the Turkish period (1571-1878). In 1879 the British administration introduced a series of Forest Laws that regulated goat grazing inside 'state forests', banned the collection of wood and reduced the amount of forest cleared for vineyards. In 1888 the Goat Law was introduced that banned importation of goats into Cyprus, and in 1913 it called for elimination of goats from village lands.[4]

These laws were partly enacted to protect the endemic mouflon (as the 'wild' sheep of Cyprus is commonly called), one of the animal species that became naturalised in antiquity and survived into modern times. Due to human exploitation and habitat destruction, by 1878 the numbers of mouflon decreased and the animals were confined to the southern mountain range in Troodos and Paphos forests (a mountainous area of 620,000 ha in the north-western part of the Troodos mountain range). In 1883 the mouflon population of the Paphos Forest dwindled to 25 animals. In 1937 the Troodos herd was exterminated and the Paphos herd was reduced to 15 mouflons only.[5] The survival of the species hung in the balance.

As these examples show, the division between 'wild' and domestic animals in Cyprus has been permeable for thousands of years, and that remains the case in modern times. The categories remain fluid and the cyclical pattern of human intervention continues.

---

4   Muhly 1986, 46; Thirgood 1987, 63–69; 1988; Michaelidou & Decker 2002, 293.
5   Anon after 1970; Michaelidou & Decker 2002, 293.

## Modern dynamics

The contemporary anthropocenic interdependencies in Cyprus are shaped by various socio-economic and political forces including modernisation of agriculture, urbanisation, globalisation, political conflict and financial crises.

Modernisation of agricultural practices in lowland areas, and declining farm profitability in remote and mountainous lands unsuitable for mechanised farming techniques led to increased animal and land abandonment throughout the 20th century in the Mediterranean. Since the 1950s, this process has been further influenced by the movement of people from the countryside to urban areas and abroad, stimulated by the development of the city-based industries and the guest-worker arrangement with Germany (Plieninger et al. 2013, 2).

The changing agricultural practices have affected donkeys in particular. As recently as the 1930s and 1940s, the number of donkeys and mules in Cyprus was 50,000 (Meyer 1962, 29, table 5). But from the mid-20th century onwards the number of donkeys has been dwindling, replaced by tractors, utility trucks and SUVs. When not needed, the animals are simply abandoned and occasionally mistreated, and depend on charities. Mary and Patrick Skinner, members of the 60,000-strong group of British expatriates living in Cyprus, established one of the first charities, The Friends of the Cypriot Donkey, in 1994. From six animals, the sanctuary, based in small village of Vouni, grew to house 37 in 1996, 57 in 1997, and more than 100 in 2004. In 2007 Mary and Patrick retired and the organisation was taken over by the Devon-based The Donkey Sanctuary and in 2013 it was a home to 120 donkeys.[6] According to the 2002 census carried out in southern Cyprus, only 2175 donkeys were left: 1700 working in rural areas, 300 in tourism, and 175 in sanctuaries (Kugler et al. 2008, 22).

The animal and land abandonment in Cyprus has also been caused by political conflict. Established in July 1974, following the Turkish invasion, the United Nations Buffer Zone in Cyprus (also known as the Green Line) runs for about 180 km, dividing the island and the capital

---

6   Marczewska 2005, 472; see also The Donkey Sanctuary. Retrieved on 23 April 2015 from http://www.donkeysanctuarycyprus.org/en/about-us.

city Nicosia/Lefkosia into two parts. The zone, with width varying from 3.5 m to 7.5 km, has become an accidental/involuntary wildlife sanctuary with a number of rare indigenous plants. Apart from a population of mouflon (200 individuals in 2009), packs of feral dogs roam the area and it is visited by a large number of migratory birds. Depending on the outcome of peacekeeping talks, the future of the zone as a sanctuary is threatened, as the establishment of a national park is unlikely (Simonsen 2009).

In addition to the Green Line, the Karpas Peninsula in the northeast of Cyprus became home to feral donkeys. The animals were left behind during the exodus of Greek Cypriots in 1974, and benefited from the subsequent closure of the area, placed under Turkish military control until the army left in 1983. (Interestingly, pigs left behind failed to establish a feral population [Reese 1992a, 16].) Once the army left, that area was declared to be a national park. In 1997 the park was extended and in 1998 an area of 155 square km of the Karpas Peninsula was finally declared the first and only national park in the Turkish Republic of Northern Cyprus which theoretically should limit urban development in the area, though in practice 'it may make little difference' (Warner 2006).

The tip of the peninsula has been fenced-off to keep donkeys contained, though animals have been found outside. In lean years they have been known to cause damage to crops and fields in the area, raising demands for the removal of the population, if not by physical extermination, then by catching and by domesticating them or exporting them to Turkey (Durduran & Gurkan nd; Hadjicostis 2009). Moreover, researchers raised concern about the 'negative impacts on vegetation and native species'. The survey carried out in 2003 estimated that the population of donkeys had grown to 800 in the entire 132.5 square km though it was concentrated within the fenced-off area. The average estimated density was 6.7 donkeys per square km.[7]

Apart from feral donkeys, Cyprus has large stray cat and dog populations. While the exact numbers of these animals are unknown, a couple of reports give indication of the massive scale of the problem. As

---

7   Hamrick et al. 2005, 108; an earlier survey carried out in December 1993 – January 1994 estimated the population of donkeys to be 337 (Reid et al. 1997).

part of the stray dog elimination program carried out between 1971 and 1984 and aimed at the elimination of echinococcosis, a parasitic disease caused by *Echinococcus granulosus* (also known as dog tapeworm) 82,984 stray dogs were destroyed. The remaining dog population was estimated at 19,955 dogs in 1984 (Polydorou 1985, 336).

A 2008 report on stray cats in Nicosia, the largest cosmopolitan area of Cyprus, estimated that there are 700 to 1000 cats per square km.[8] Just for comparison, cat densities recorded at three rubbish dumps (places of high densities due to easy food availability) on the outskirts of Canberra were assessed to be 90 cats per square km at Mac's Reef Road, 38 cats per square km at Belconnen, and 19 cats per square km at Mugga Lane (Denny & Dickman 2010, 12). As Cypriot resident Dan Rhoads notes, 'cat populations are thriving in every town and small city in Cyprus. Roadkill is just as pervasive here as it is in the United States, but unlike the USA, 99 out of a 100 carstrikes involves a cat.'[9] The impact of cat population on birdlife in Cyprus is unknown but it is most likely significant. Following a systematic review of the impact of free-ranging domestic cats on wildlife in the USA, Loss et al. estimated that they, especially unowned animals, 'kill 1.4–3.7 billion birds and 6.9–20.7 billion mammals annually' in the USA and suggest that 'free-ranging cats cause substantially greater wildlife mortality than previously thought and are likely the single greatest source of anthropogenic mortality for US birds and mammals' (Loss et al. 2013, 1396).

The stray animal populations have been typically dealt with by shooting and by poisoning, but a combination of efforts by animal welfare groups, many started by British expatriates living in Cyprus, and the need to comply with the animal laws of the European Union following 2004, is slowly changing the Cypriot laws and local attitudes. For example, a dog law introduced on 1 October 2004 aims to reduce stray

---

8   Peter M Heise-Pavlov & Eleftherious Hadjisterkotis, *The cat* (Felis sylvestris catus L.) *in the city of Lefkosia: public health problems and options for management,* 2nd edn. Nicosia: Ministry of Interior 2008 quoted by Dan Rhoads. Retrieved on 18 May 2015 from https://migration.wordpress.com/2009/09/10/feral-cats-and-birds/.
9   Retrieved on 23 April 2015 from https://migration.wordpress.com/2009/09/10/feral-cats-and-birds/.

dog populations by 'controlling indiscriminate breeding, dog abandonment and theft' (Voslářvá & Passantino 2012, 98).

Although the law changed, its implementation has been lagging behind. Societal values and cultural attitudes evolve slowly, judging by
the number of complaints in the online media, especially from British
holidaymakers who find the mistreatment of dogs in Cyprus deeply
upsetting.[10] Moreover, the recent financial crisis in Cyprus has put a
stop to many initiatives. For example, the cuts to the government-
funded spaying/neutering programs resulted in 'the mushrooming cat
population'.[11]

The various socio-economic and political forces, and cultural attitudes are also affecting animal conservation issues.


## Conservation conundrums

The majority of terrestrial non-volant mammals of Cyprus, such as the
endemic mouflon, fox, hare and donkey, include anthropophorous taxa
of ancient origin. Are these 'native' species and should they be protected
by law? The traditional classification of animal species into dichotomies
of domestic/wild and alien/native remains 'one of the organising principles of conservation biology and restoration ecology' (Warren 2007,
428) as a way of maintaining boundaries between nature (represented
by native and wild species) and culture (represented by domestic and
alien species) (Milton 2013, 112).

Typically 'native species are those which have autocolonized an
area since a selected time in the past ... and alien species are those
which have been introduced by humans, intentionally or otherwise'
(Warren 2007, 428). Rees suggested that time could be used as a criterion for establishing a 'native' status of an animal species. Time is also a

---

10   See, for example, Organised campaign against animal cruelty in Cyprus
begins, 25 January 2014. Retrieved on 23 April 2015 from:
http://www.cyprusexpat.co.uk/blog/read/id:4381/
organised-campaign-against-animal-cruelty-in-cyprus-begins.
11   Invasion of stray cats as spaying programmes dry up, *Cyprus Mail*, 20 August
2013. Retrieved on 23 April 2015 from http://cyprus-mail.com/2013/08/18/
invasion-of-stray-cats-as-spaying-programmes-dry-up/.

useful point of reference when deciding on 'how far back in time should we look for historical evidence of the former natural occurrence of a species' and 'how far should we go in attempting to reintroduce locally extinct species?' (Rees 2001, 218). In view of the transgressive character of the animals in Cyprus, another question comes to the fore: how long does is take for a feral animal population living independently from humans to be considered 'wild'?

While the terms 'feral' and 'wild' were originally synonymous and are still used interchangeably as antonyms for something that is 'tame' or 'domesticated', they carry contrasting connotations. The adjective 'feral' denotes a domestic animal that escaped and established a sustainable population independent from humans. It is often an introduced species alien to the landscape that it now inhabits and is frequently considered a pest. As is the case with a 'stray' animal. These three categories: stray, feral and wild can be seen in a continuum of varying reliance on human environment: from stray animals (recent or temporary domestic escapees, or abandoned pets, that rely on humans for food and shelter) to feral animals (descendants of once-domestic animals that breed and survive in the 'wild' independently) to wild animals (those that were never domesticated, or reverted to the wild state 'a long time ago'). The length of time required for the feral animal to become naturalised and to be considered 'wild' varies and depends on cultural values, perception of specific animal species and other issues. Various stakeholders in a society can see the same group of animal in conflicting light depending on their relationship to animals and landscape. For example, in Australia feral brumbies in the Snowy Mountains are classified as pests by the New South Wales National Park authorities, but they attract wide public support as an important part of cultural heritage and representatives of a charismatic species (Silmalis 2013).

In the context of the Mediterranean, the length of time required for an alien animal to become native, and feral to become wild, has been conveniently formalised. The Red List of Threatened Species™ released by the International Union for Conservation of Nature included 'all terrestrial mammal species native to the Mediterranean or naturalized since before 1500 A.D.' (Temple & Cuttelod 2009, vi). The report recognises that the presence of many of the animals resulted from human intervention and it recommends that:

any conservation strategy aimed at maintaining biodiversity and its evolutionary potential takes into account the history (including recent history) of the regional biota, and makes an effort: (1) to identify and direct attention towards ancient endemic species that escaped previous extinction events and are the repository of unique phylogenetic information; and (2) to strike an appropriate balance between conserving large, charismatic mammals (that may in some cases be relatively recent additions to the regional fauna) and protecting other forms of native biodiversity. (Temple & Cuttelod 2009, 19)

The conservation efforts in the Mediterranean have been skewed towards the protection of a small number of island endemics – 'barely diagnosable taxa of recent anthropochorous origin that are presently restricted to well-determined geographical areas' (Gippoliti & Amori 2006, 42). As shown before these animals come from introduced animals, domestic or wild, or a mixture of different stock reflecting the multiple waves of introductions. For Spartaco Gippoliti and Giovanni Amori their value is 'low compared to native taxa, unless they are the only descendants of now extinct continental populations', and they suggest that 'all known anthropochorous taxa should be excluded from international and possibly national protective legislation'. Instead the priority should be given to 'the few extant palaeoendemic insular mammals' and endemic plants (Gippoliti & Amori 2006, 42).

In contrast to Gippoliti and Amori, Marco Masseti considers such a 'purist' approach to conservation unnecessary and potentially detrimental to the island biodiversity as the natural ecosystems of the Mediterranean had been 'irretrievably destroyed' thousands of years ago. He suggests that 'the anthropochorous mammalian populations of certified ancient origin' represent important 'historic documents' and should be 'considered as authentic "cultural heritage" '(Masseti 2009a, 189; 2009b, 303).

While this definition would explain the special status of the Cypriot mouflon, it leaves the feral donkey in a precarious position as a relatively recent escapee. Interestingly, feral donkeys 'have been broadly adopted as a cuddly symbol of the island's agrarian past' (Hadjicostis 2009) and even a symbol of Cypriot nationality.[12] This may impact their long-term fate. Rauf Denktaş (former member of the Turkish Cypriot nationalist leadership until 1974 and former president of the Turk-ish Republic of Northern Cyprus until April 2005) is reported to have said that 'the only Cypriot living in Cyprus is the Cyprus donkey'.[13] While the idea of exporting the feral donkey population to Turkey was rejected, there is local support for rounding up 'sustainable' donkey populations, placing them in smaller animal sanctuaries or ecotourism enterprises, such as petting zoos or farms (Hadjicostis 2009).

Finally, it has been suggested that the alien/native framework should be replaced with a 'degradation criterion' (see discussion in Warren 2007, 437), and animal species should be judged 'not on their origins but simply on the basis of whether they are ecologically, eco-nomically, or otherwise problematic' (Simberloff 2012, 11). In the case of Cyprus, conservation efforts should focus on managing the species that cause harm in the environment, such as rats (black rat, *Rattus rat-tus* and brown rat, *Rattus norvegicus* Berkenhout 1769), stray dogs and cats.

From this overview of conservation issues, it can be seen that the question of which animals should be protected is complex in view of the historicity of the animal introductions to Cyprus. The preference for large, 'charismatic' animals means that smaller mammals (but also reptiles and amphibians) that can claim Pleistocene origin receive little attention. Even worse, these animals are often perceived as pests and are targets of purposeful eradication. Such is the fate of snakes or bats for example.

Remains of at least three species of snakes were present in Akrotiri Aetokremnos: the blunt-nosed viper (*Vipera lebetina*), the Cypriot grass snake (*Natrix natrix cypriaca*) and Persian whip snake (*Coluber jugularis*). Apart from reptiles and amphibians (remains of Green toad, *Bufo viridis*, were also found in Akrotiri Aetokremnos), only two small extant mammals seem to be relics of indigenous Pleistocene taxa: the Cypriot white-toothed shrew and the Cypriot mouse. These smaller animals require specific conservation measures focused on raising of

---

12  Donkey campaign unites Cypriots, BBC News, 22 April 2008. Accessed on 18 May 2015 from http://news.bbc.co.uk/2/hi/europe/7361353.stm.
13  Cited in the Turkish Cypriot daily *Ortam*, 13 November 1995 (Ramm 2006, 527).

public awareness of their diversity and ecological importance in order to improve their 'pest' image. The threats are chiefly linked with the loss and the degradation of habitat as a result of agricultural intensification and land abandonment, the prevention of which has been linked with more sustainable agricultural practices (Temple & Cuttelod 2009, vii).

In contrast to small mammals, the conservation of larger mammals requires better management of protected areas, improved implementation of existing laws, and regulations controlling hunting, and development and implementation of species-specific management plans (Temple & Cuttelod 2009, vii). For large herbivores, the maintenance of grazing systems is important, as well as the understanding and support of local communities. As mentioned before, this support is often lacking where people struggle with unprofitable agriculture, lack of educational and employment opportunities, and resulting declining populations in communities.

Sometimes the loss of traditional means of survival is unavoidable (as in the case of goat herding). Michaelidou and Decker postulate however that 'people who derive their livelihoods from nature often have a stake in conserving local biodiversity and may become important allies to conservation efforts' (Michaelidou & Decker 2002, 291). In particular, they argue that traditional land use systems with fluctuating regimes and intensity may have benefited biological diversity in the region. According to Jacques Blondel, 'moderate grazing intensity has the potential to maximise species diversity and optimise ecosystem productivity' (Blondel 2006, 723, 726).

Even though Cyprus has had a long history of animal introductions, I agree with Masseti who says, 'in view of the vulnerability of the insular ecosystems it is critically important to prevent further introductions' (Masseti 2009a, 169), including those that could theoretically be construed as native restoration such as wild boar or deer. The Convention on the Conservation of European Wildlife and Natural Habitats (1979), also known as the Berne Convention, was the first treaty that encouraged the reintroduction of native species as a method of conservation (later replaced by 'an obligation simply to study the desirability of reintroducing them'). As Rees points out, neither the Berne Convention nor the 1992 Convention on Biological Diversity defined what a native species was, which was likely to lead to inconsistencies, conflicting interpretations of international law and irresponsible (re)introductions (Rees 2001, 216–18). This is what has happened in Cyprus.

Deer were reintroduced in 1980 as a result of the Cypriot government's initiative to breed the Mesopotamian fallow deer. Unfortunately, two female and one male shipped from a zoo in Switzerland were of the wrong species (European fallow deer). The attempts to send the animals back were unsuccessful and 20 years later, in 2002, 50 to 60 European fallow deer were housed near Stavros tis Psokas. They could not have been released into the wild due to their rapid reproduction rate and the associated danger to the environment and the populations of endemic mouflon. It had been suggested that the deer should be slaughtered, but later the Forestry Department decided to give the animals away to zoological gardens or parks for private keeping (Saoulli 2002).

In 1990, a farmer imported five wild boars from Greece for game farming. A group of animals was released in 1994 in Limassol Forest, and a year later another group was released in the Troodos National Forest Park. In 1997 the Cypriot government decided to eradicate the population to prevent possible environmental destruction and the danger of transmitting diseases to livestock. By 2001/02, the population of wild boar between Troodos and Plastres was estimated to be 80 animals, but surveys carried out in 2004 and 2005 failed to observe any animals. The rapid extermination was linked to inbreeding associated with reduced reproductive fitness, food shortage and the introduction of improved ammunition (Hadjisterkotis & Heise-Pavlov 2006, 213–14).

The illegal introduction of wild boar aimed at enriching local game fauna for hunting. Surprisingly, keeping in mind the paucity of terrestrial wildlife, hunting remains a popular activity in Cyprus with 45,000 hunters reported in 2007. With 6.4 percent of Cypriots (25 percent of men aged more than 18) engaged in hunting, Cyprus has the second highest per capita population of hunters in Europe.[14] Hare (*Lepus europeus*) is the main terrestrial game species with approximately 50,000 animals killed each year.[15] But it is the birds such as

---

14  Following Ireland with 8.9 percent. Compare with 1.4 percent in the UK and 0.3 percent in Poland (Anon 2005; Dickson et al. 2009, 32).

chukar partridge (*Alectoris chukar*), black francolin (*Francolinus francolinus*), woodpigeon (*Columba palumbus*), turtle-dove (*Streptopelia turtur*), skylark (*Alauda arvensis*) and thrush species (mainly *Turdus philomelos* and *T. merula*) that are the chief hunting spoils (Nicolaos Kassinis in Anon 2005). Experts estimate that song thrushes are killed in Cyprus in numbers up to 800,000 a day and about 20 million or more a year (and turned into a Cypriot traditional dish known as *ambelopoulia*), raising international concerns about the future of these species (Chapman 2002).

These concerns are well grounded. Human predation remains one of the chief drivers of animal extinctions (Sodhi et al. 2009). They can be exemplified in the case of the Cypriot mouflon. As mentioned above, in 1937 the population of endemic mouflon comprised only 15 individuals. In 1939 the Paphos Forest was declared a Permanent Game Reserve, which banned all hunting. Since these changes were introduced, the numbers of mouflon have steadily increased.[16] In 2012 the population of Cypriot mouflon in the Paphos Forest was estimated to be 3000 individuals.[17] Although a conservation success story, the increased number of mouflon meant that conflict with local people was unavoidable. During a survey in 2000, members of local village communities, while generally supporting the need for the conservation of mouflon, 'expressed their disapproval and indignation for the agricultural damages caused by mouflon to local vineyards and fruit trees' (Michaelidou & Decker 2002, 296). A similar conflict between 'wild' and rural space is present in the Karpas Peninsula. Moreover, both the Karpas Peninsula and Paphos Forest are threatened with habitat degradation in a similar way to Yellowstone Park, USA, where the lack of

---

15   Retrieved on 23 April 2015 from http://www.cypenv.info/cypnat/files/mammals.aspx .

16   Anon after 1970; Michaelidou & Decker 2002, 293. The Zoological Garden of Limassol was involved in a breeding program of the Cyprus mouflon since 1955, but in 1993 the program was assessed as a failure. The surplus animals were reintroduced to a free-ranging population in the Paphos Forest without proper veterinary examination. Moreover, they were of reduced viability (due to inbreeding) and could have potentially endangered the wild population as carriers of contagious diseases (Hadjisterkotis & Bider 1993).

17   Anon after 1970; Michaelidou & Decker 2002, 293; Barbanera et al. 2012, 671.

human interference in bison management has resulted in serious over-grazing (Braje 2011, 73).

## Conclusions

As a result of human intervention Cyprus was transformed from an is-land with limited biodiversity into one with a much richer non-volant terrestrial animal world. Although predominantly anthropogenic in its origin, the mammalian biodiversity of Cyprus has been fragile and many animal species have remained vulnerable to extinction 'due to a limited source of conspecific individuals to replenish dwindling pop-ulations' (Grayson 2001, 34). Cyprus is an exemplary case of anthro-pocenic interdependencies. Since antiquity to the present, the ecolog-ical assemblage has been closely intertwined with the changing eco-nomic, social and political situation, and it has remained a place of disturbance and change.

The Cypriot 'wild' mammals such as hare, fox and mouflon (which humans introduced in antiquity) are considered to be native to the island and rely on human protection and management for their long-term survival. The mouflon and the fox are protected in Cyprus, but even hare with its high reproductive capacity is not safe from the im-pact of habitat degradation and loss, drought, chemicals used in agri-culture, stray dogs and overhunting by humans. In order to maintain a viable population for hunting, the breeding and release of captive hares have been necessary (Marios 2009, 29–30). In January 2014 the Cypriot partliament was considering changing the legal status of red foxes from a protected to game animal, on the basis of the potential consequences of the population of foxes on the farming sector and wildlife. As the data about the increase in the population of foxes are conflicting, it seems to be another case of self-interested and poten-tially misguided attempt to protect the interests of farmers and hunters. Misguided, because according to the Initiative for the Protection of the Cyprus Fox, foxes actually have a positive impact on agriculture as the natural predators of rats and crickets.[18]

The representation of the feral donkey as an 'endangered wild ani-mal', a 'symbol of the island' and the 'only Cypriot living in Cyprus' in the public discourse shows how quickly a species can change its status from domestic to feral to wild in favourable ecological and cultural circumstances. And although local communities regard these animal species with pride and sentiment, they become 'matter out-of-place' (Mary Douglas after Milton 2013) once they cross the physical boundaries between the 'wilderness' and rural areas, and affect the farmers' livelihood.

While feral donkeys are becoming an accepted part of the 'wilderness', provided that they do not leave the fenced-off area, the stray populations of dogs and cats that overrun Cypriot town and villages remain perceived as major pests. Although their impact on other species, especially birds, has not been researched in detail, for years, if not centuries, they have continued to pose a serious public health, socioeconomic and animal welfare problems.

As people have been critical in the creation of biodiversity, they remain a vital agent in its preservation. But taking responsibility for the animal world on a small island has been littered with examples of fixing one problem by introducing another, which questions the capacity of humankind to effectively mitigate anthropocenic damages. Too many decisions, in Cyprus and elsewhere, seem to be governed primarily by economic concerns, 'not in my backyard' opposition, tradition or a short-sighted view of the environmental issues. The cultural, political and legal contexts might be slowly changing, but the societal values and individual behaviour are too often grounded in a purely utilitarian approach to animals. All these forces combined are driving the shape of human–animal relations in Cyprus and will determine the future of the island ecosystem in the Anthropocene.

I agree with Donald Grayson who writes, 'Pleistocene biotas cannot be recreated no matter how hard we try . . . The clear implication is that it is the active management of biotic landscapes, not the exclusion of human influence, that is needed to accomplish these goals' (Grayson 2001, 46). While the Pleistocene ecosystem cannot be recreated in Cyprus, the archaeological data could be used to inform the conser-

---

18    Protect the Cyprus fox, *In Cyprus*, 19 February 2014. Retrieved 23 April 2015 from http://web.archive.org/web/20140307031402/http://incyprus.philenews.com/en-gb/Top-Stories-News/4342/40004/protect-the-cyprus-fox.

vation decision and to educate the public in order to identify problem areas.

The permeability of boundaries between the native/alien and wild/ domestic dichotomies in Cyprus demonstrates the shortcomings of the current conservation principles. Conservationists in Cyprus and else-where, however, are likely to cling to these distinctions as a way to justify their work, relying on the preservation of boundaries between nature and culture, and between human and non-human processes. As Kay Milton (2013, 113) shows, 'the very project of nature conservation is contradictory, since it seeks to conserve what is natural through un-natural means (human agency)'. As the world ecosystems are increas-ingly altered by human activities in the Anthropocene, the concepts of nature and wilderness as places free of human impact are becoming ob-solete, and these dichotomies are becoming irrelevant. Moreover, the concept of 'taking responsibility' should also extend to domestic and feral animals.

Taking responsibility for environment in the Anthropocene re-quires accepting the human-altered landscapes and ecosystems and making them as sustainable as possible. Environmentalists, conser-vationists and local communities need to take transformative action; action that is informed by scientific knowledge, and cultural, social and political sciences, including archaeology. There is a need for other approaches to the animal conservation and management issues; ap-proaches that look at the impact of individual species on its ecosystem instead of its biological lineage, mitigate global processes of species ex-tinction and invasion, and aim to sustain the existing biodiversity as a matter of priority.

## Works cited

Anderson K (1998). Animals, science, and spectacle in the city. In JR Wolch & J Emel (Eds). *Animal geographies: place, politics, and identity in the nature–culture borderlands* (pp27–50). London & New York: Verso.

Anon (after 1970). The Cyprus mouflon, *Ovis ammon orientalis* (Cyprus variety). Cyprus American Archaeological Research Institute, Nicosia, CS FAU CYP Acc.742, 4pp.

Anon (2005). Toward a common monitoring scheme of hunting bag in the European Union. AEWA/TC6 Inf. 6.11, Conference Proceedings. Final discussion paper and agenda (pp1–10). Sustainable Hunting Initiative, Technical Workshop, Station Biologique de la Tour du Valat, 16–17 March 2005.

Balter M (2013). Archaeologists say the 'Anthropocene' is here – but it began long ago. *Science* 340(6130): 261–62.

Barbanera F, Guerrini M, Beccani C, Forcina G, Anayiotos P & P Panayides (2012). Conservation of endemic and threatened wildlife: molecular forensic DNA against poaching of the Cypriot mouflon (*Ovis orientalis ophion*, *Bovidae*). *Forensic Science International: Genetics* 6(5): 671–75.

Blondel J (2006). The 'design' of Mediterranean landscapes: a millennial story of humans and ecological systems during the historic period. *Human Ecology* 34(5): 713–29.

Boekschoten GJ & Sondaar PY (1972). *On the fossil mammalia of Cyprus I–II.* Proceedings of the Koninklijke Nederlandse Akademie van Wetenschappen. Amsterdam & London: North-Holland Publishing Company.

Bowker GC & Star SL (1999). *Sorting things out: classification and its consequences.* Cambridge, MA: MIT Press.

Braje T (2011). The human–animal experience in deep historical perspective. In C Blazina, G Boyra & D Shen-Miller (Eds). *The psychology of the human–animal bond* (pp63–80). New York: Springer.

Budiansky S (1994). *The covenant of the wild: why animals chose domestication.* London: Weidenfeld & Nicholson.

Chapman J (2002). Hunters on Cyprus slaughter 800,000 thrushes in a day. *Daily Mail*, 9 July.

Croft P (1991). The animal remains. In E Peltenburg. *Lemba Archaeological Project. Volume II.2. A ceremonial area at Kissonerga* (pp73–74). Göteborg: Paul Åströms Förlag.

Croft P (2002). Game management in early prehistoric Cyprus. *Zeitschrift für Jagdwissenschaft* 48(Supplement): 172–79.

Cucchi T, Orth A, Auffray J-C, Renaud S, Fabre L, Catalan J, Hadjisterkotis E, Bonhomme F & J-D Vigne (2006). A new endemic species of the subgenus *Mus* (Rodentia, Mammalia) on the island of Cyprus. *Zootaxa* 1241: 1–36.

Denny EA & Dickman CR (2010). Review of cat ecology and management strategies in Australia: a report for the Invasive Animals Cooperative Research Centre. Sydney: Institute of Wildlife Research, School of Biological Sciences, University of Sydney.

Diamond JM (1992). Twilight of the pygmy hippos. *Nature* 359: 15.

Dickson B, Hutton J & Adams B (Eds) (2009). *Recreational hunting, conservation and rural livelihoods: science and practice.* Oxford, UK: Wiley-Blackwell.

Driscoll CA, Macdonald DW & O'Brien SJ (2009). From wild animals to domestic pets, an evolutionary view of domestication. *Proceedings of the National Academy of Sciences* 106(Supplement 1): 9971–78.

Dubey S, Cosson JF, Magnanou E, Vohralík V, Benda P, Frynta D, Hutterer R, Vogel V & P Vogel (2007). Mediterranean populations of the lesser white-toothed shrew (*Crocidura suaveolens* group): an unexpected puzzle of Pleistocene survivors and prehistoric introductions. *Molecular Ecology* 16(16): 3438–52.

Durduran T & Gurkan C (nd). Our donkeys – Esseklerimiz. Text of a high-school video documentary. Retrieved 20 August 2013, from www.stwing.upenn.edu/~durduran/donkey2.txt.

Gippoliti S & Amori G (2006). Ancient introductions of mammals in the Mediterranean basin and their implications for conservation. *Mammal Review* 36(1): 37–48.

Grayson DK (2001). The archaeological record of human impacts on animal populations. *Journal of World Prehistory* 15(1): 1–68.

Hadjicostis M (2009). Eeyore plague: wild donkeys overrun Cyprus villages. *Independent*, 31 May. Retrieved on 30 April 2015 from http://www.independent.co.uk/environment/nature/ eeyore-plague-wild-donkeys-overrun-cyprus-villages-1693518.html.

Hadjisterkotis ES & Bider JR (1993). Reproduction of Cyprus mouflon *Ovis gmelini ophion* in captivity and in the wild. *International Zoo Yearbook* 32(1): 125–31.

Hadjisterkotis ES & Heise-Pavlov P (2006). The failure of the introduction of wild boar *Sus scrofa* in the island of Cyprus: a case study. *European Journal of Wildlife Research* 52(3): 213–15.

Hamrick R, Pirgalioglu T, Gunduz S & J Carroll (2005). Feral donkey *Equus asinus* populations on the Karpaz peninsula, Cyprus. *European Journal of Wildlife Research* 51(2): 108–16.

Held SO (1989). Colonization cycles on Cyprus I: the biogeographic and palaeontological foundations of early prehistoric settlement. *Report of the Department of Antiquities, Cyprus*: 7–28.

Held SO (1992). Colonization and extinction on early prehistoric Cyprus. *Acts of an International Congress on Cypriote Archaeology held in Göteborg on 22–24 August 1991. Part 2* (pp104–64). Jonsered, Sweden: Paul Åströms Förlag.

Kugler W, Grunenfelder H-P & Broxham E (2008). Donkey breeds in Europe: inventory, description, need for action, conservation. St Gallen, Switzerland: Monitoring Institute for Rare Breeds and Seeds in Europe. Report 2007/2008.

Loss SR, Will T & Marra PP (2013). The impact of free-ranging domestic cats on wildlife of the United States. *Nature Communications* 4: 1396.

Marczewska A (2005). Animals in ancient Cyprus. PhD thesis, Sydney: Department of Archaeology, University of Sydney.

Marios M (2009). Analysis of Cyprus hare population from breeding and hunting point of view. Thesis, Budapest: Faculty of Veterinary Science, Shent István University.

Masseti M (1998). Holocene endemic and anthropochorous wild mammals of the Mediterranean islands. *Anthropozoologica* 28: 3–20.

Masseti M (2009a). Mammals of the Mediterranean islands: homogenisation and the loss of biodiversity. *Mammalia* 73(3): 169–202.

Masseti M (2009b). A possible approach to the 'conservation' of the mammalian populations of ancient anthropochorous origin of the Mediterranean islands. *Folia Zoologica* 58(3): 303–08.

Meyer AJ (1962). *The economy of Cyprus*. Cambridge, MA: Harvard University Press.

Michaelidou M & DJ Decker (2002). Challenges and opportunities facing wildlife conservation and cultural sustainability in the Paphos Forest, Cyprus: historical overview and contemporary perspective. *Zeitschrift für Jagdwissenschaft* 48(1): 291–300.

Milton K (2013). Ducks out of water: nature conservation and boundary maintenance. In H Callan, B Street & S Underdown. *Introductory readings in anthropology* (pp105–13). New York: Berghahn Books.

Mrva-Montoya A (2013). Learning from dead animals: horse sacrifice in ancient Salamis and the Hellenisation of Cyprus. In J Johnston & F Probyn-Rapsey (Eds). *Animal death* (pp169–88). Sydney: Sydney University Press.

Muhly JD (1986). The role of Cyprus in the economy of the Eastern Mediterranean during the second millennium BC. In V Karageorghis (Ed). *Acts of the International Archaeological Symposium 'Cyprus Between the Orient and the Occident', Nicosia, 8–14 September 1985* (pp45–60). Nicosia: Department of Antiquities, Cyprus.

Peltenburg E (1998). The character and evolution of settlements at Kissonerga. In *Lemba Archaeological Project. Volume II.1A. Excavations at Kissonerga-Mosphilia, 1979–1992* (pp233–60). Jonsered, Sweden: Paul Åströms Förlag.

Plieninger T, Gaertner M, Hui C & L Huntsinger (2013). Does land abandonment decrease species richness and abundance of plants and animals in Mediterranean pastures, arable lands and permanent croplands? *Environmental Evidence* 2(1): 1–7.

Polydorou K (1985). The eradication of serious animal diseases/zoonoses in Cyprus – echinococcosis, brucellosis and anthrax. *International Symposia on Veterinary Epidemiology and Economics Proceedings, ISVEE 4: Proceedings of the 4th International Symposium on Veterinary Epidemiology and Economics, Singapore, Establishing priorities and evaluation of international and national programs session* (pp335–37). International Symposia on Veterinary Epidemiology and Economics.

Ramm C (2006). Assessing transnational re-negotiation in the post-1974 Turkish Cypriot community: 'Cyprus donkeys', 'black beards' and the 'EU carrot'. *Southeast European and Black Sea Studies* 6(4): 523–42.

Rees PA (2001). Is there a legal obligation to reintroduce animal species into their former habitats? *Oryx* 35(03): 216–23.

Reese DS (1992a). Background to the past and present wild mammals of Cyprus. Cyprus American Archaeological Research Institute, Nicosia, CS FAU REE Acc.3213, pp12–18.

Reese DS (1992b). Tale of the pygmy hippo. *Archaeology* 6: 50–53.

Reese DS (1993). Folklore and fossil bones: the pygmy mammals of Cyprus. *Terra Nova* 42: 319–21.

Reese DS (1995). *The Pleistocene vertebrate sites and fauna of Cyprus*. Nicosia, Cyprus: Ministry of Agriculture, Natural Resources and Environment.

Reid SWJ, Godley BJ, Henderson SM, Lawrie GJ, Lloyd D, Small K, Swannie N & RL Thomas (1997). Ecology and behaviour of the feral donkey, *Equus asinus*, population of the Karpas Peninsula, Northern Cyprus. *Zoology in the Middle East* 14(1): 27–36.

Rick TC, Kirch PV, Erlandson JM & Fitzpatrick SM (2013). Archaeology, deep history, and the human transformation of island ecosystems. *Anthropocene* 4: 33–45.

Rowley-Conwy P, Albarella U & Dobney K (2012). Distinguishing wild boar from domestic pigs in prehistory: a review of approaches and recent results. *Journal of World Prehistory* 25(1): 1–44.

Ruddiman WF (2003). The anthropogenic greenhouse era began thousands of years ago. *Climatic Change* 61(3): 261–93.

Ruddiman WF (2013). The Anthropocene. *Annual Review of Earth and Planetary Sciences* 41(1): 45–68.

Saoulli A (2002). Wanted: large space for fallow deer. *Cyprus Mail*, 21 February.

Sax DF & Gaines SD (2008). Species invasions and extinction: the future of native biodiversity on islands. *Proceedings of the National Academy of Sciences* 105(Suppl 1): 11490–97.

Silmalis L (2013). Wild brumbies in Snowy Mountains may be culled with aerial shooting. *Sunday Telegraph*, 16 June. Retrieved on 30 April 2015 from http://tiny.cc/tuajzx.

Simberloff D (2012). Nature, natives, nativism, and management: worldviews underlying controversies in invasion biology. *Environmental Ethics* 34(1): 5–25.

Simmons AH (1992). Akrotiri-Aetokremnos and Early Cypriot prehistory. In P Åström (Ed). *Acta Cypria 2: Acts of an International Congress on Cypriote Archaeology Held in Göteborg on 22–24 August 1991* (pp348–55). Jonsered, Sweden: Paul Åström Förlag.

Simmons AH (1999). *Faunal extinction in an island society: pygmy hippopotamus hunters of Cyprus.* New York: Kluwer Academic/Plenum Publishers.

Simmons AH (2001). The first humans and last pygmy hippopotami of Cyprus. In S Swiny (Ed). *The earliest prehistory of Cyprus: from colonization to exploitation.* Vol. 2 (pp1–18). Boston, MA: American School of Oriental Research.

Simonsen SG (2009). The casualties of peace in Cyprus. *Independent*, London: 26.

Smith BD & Zeder MA (2013). The onset of the Anthropocene. *Anthropocene* 4: 8–13.

Sodhi NS, Brook BW & CJA Bradshaw (2009). Causes and consequences of species extinctions. In SA Levin (Ed). *The Princeton guide to ecology* (pp514–20). Princeton, NJ: Princeton University Press.

Solli B, Burström M, Domanska E, Edgeworth M, González-Ruibal A, Holtorf C, Lucas G, Oestigaard T, Smith L & C Witmore (2011). Some reflections on heritage and archaeology in the Anthropocene. *Norwegian Archaeological Review* 44(1): 40–88.

Sondaar PY (1991). Island mammals of the past. *Science Progress* 75: 249–64.

Sondaar PY & van der Geer AAE (2000). Mesolithic environment and animal exploitation on Cyprus and Sardinia/Corsica. In MA Mashkour, A Choyke, H Buitenhuis & F Poplin (Eds). *Archaeozoology of the Near East IV A. Proceedings of the fourth international symposium on the archaeozoology of southwestern Asia and adjacent areas* (pp67–73). Groningen, The Netherlands: Centre for Archaeological Research and Consultancy.

Steffen W, Crutzen J & JR McNeill (2007). The Anthropocene: are humans now overwhelming the great forces of Nature? *Ambio* 36(8): 614–21.

Swiny S (1988). The Pleistocene fauna of Cyprus and recent discoveries on the Akrotiri Peninsula. *Report of the Department of Antiquities, Cyprus* 1: 1–14, pl 1.

Temple HJ & Cuttelod A (2009). The status and distribution of Mediterranean mammals. *The IUCN Red List of Threatened Species™ - regional assessment.*

Gland, Switzerland & Cambridge, UK: International Union for Conservation of Nature.

Theodorou GE, Roussiakis SI, Athanassiou A, Giaourtsakis I & I Panayides (2007). A Late Pleistocene endemic genet (*Carnivora, Viverridae*) from Aghia Napa, Cyprus. *Bulletin of the Geological Society of Greece. Proceedings of the 11th International Congress, Athens, May, 2007*, XXXX: 201–08.

Thirgood JV (1987). *Cyprus: a chronicle of its forests, land, and people.* Vancouver, BC: University of British Columbia Press.

Vigne J-D (2011). The origins of animal domestication and husbandry: a major change in the history of humanity and the biosphere. *Comptes Rendus Biologies* 334(3): 171–81.

Vigne J-D, Carrère I, Briois F & Guilaine J (2011). The early process of mammal domestication in the Near East: new evidence from the Pre-Neolithic and Pre-Pottery Neolithic in Cyprus. *Current Anthropology* 52(S4): S255–S71.

Vigne J-D, Martin C, Moreau C, Comby C, Guilaine J, Briois F, Zazzo A, Willcox G, Cucchi T, Thiébault S, Carrère I, Franel Y & R Touquet (2012). First wave of cultivators spread to Cyprus at least 10,600 years ago. *Proceedings of the National Academy of Sciences of the United States of America* 109(22): 8445.

Vigne J-D, Zazzo A, Saliège J-F, Poplin F, Guilaine J & A Simmons (2009). Pre-Neolithic wild boar management and introduction to Cyprus more than 11,400 years ago. *Proceedings of the National Academy of Sciences* 106(38): 16135–38.

Voslářvá E & Passantino A (2012). Stray dog and cat laws and enforcement in Czech Republic and in Italy. *Annali dell'Istituto Superiore di Sanità* 48: 97–104.

Warner J (2006). Preservation and politics: a national park in North Cyprus. In D Harmon (Ed). *People, places, and parks. Proceedings of the 2005 George Wright Society Conference on Parks, Protected Areas, and Cultural Sites* (pp180–85). Hancock, MI: The George Wright Society.

Warren CR (2007). Perspectives on the 'alien' versus 'native' species debate: a critique of concepts, language and practice. *Progress in Human Geography* 31(4): 427–46.

Zalasiewicz J, Williams M, Steffen W & P Crutzen (2010). The new world of the Anthropocene. *Environmental Science & Technology* 44(7): 2228–31.

# 4

# Ecosystem and landscape: strategies for the Anthropocene

*Adrian Franklin*

## The objects and subjects of the Anthropocene

The Anthropocene designates the period after which humanity became an overbearing influence on the ordering of the Earth's biosphere. The trouble is, it is not clear when this happened. In coining the term Andrew Revkin (2011, 134) linked it to the concept of global warming and the tightening grip of recent anthropogenic change, suggesting that 'we were just entering an age that someday may be referred to as the Anthropocene'. Other scientists have disagreed over apt beginnings for such a momentous period, and they range wildly, from the Industrial Revolution in 18th-century Europe (Crutzen & Stoermer 2000), to Roman industrialism 2000 years ago (Certini & Scalenghe 2014), to the beginnings of agriculture in the Neolithic, c. 8000 years ago (Ruddiman 2003), and even as early as the mid-Pleistocene, when human hunters began to cause substantial extinctions, the loss of biodiversity and to

Franklin A (2015). Ecosystem and landscape: strategies for the Anthropocene. In Human Animal Research Network Editorial Collective (Eds). *Animals in the Anthropocene: critical perspectives on non-human futures.* Sydney: Sydney University Press.

use fire torch technologies (in Australia from 39,000 years ago) (Pyne 2001; Prideaux 2012).

If we concede that all of these make a reasonable case then the concept 'Anthropocene' no longer differentiates human history into those times and modes of production that were environmentally benign or were sustainable and those that were not. Since they cover almost all of human history the concept no longer offers the same hope for recovery or salvation. Notwithstanding Nigel Clark's (2011) warning about our '*susceptibility* to the earth's eventfulness', it suggests that for *Homo sapiens* history is disturbance; that we are planet-changers by nature. As an ecologically reflexive species, humanity lacks a species-specific niche or consistent behaviour in global ecosystems, their only shared characteristic being the capacity for significant transition and change. So here the idea of the Anthropocene (usefully) opens up our thinking about human impact and human association with animals through longer time frames than we are used to deploying.

Its expanded time scale radically alters a long-held and romantic notion that before industrialisation and the spread of capitalist domination, humans lived in a largely sustainable relationship with their environments. As a discourse emerging from the particularities of environmental thinking in the 1950s and 1960s, it retrofitted pre-industrial peoples as a 'model species' into the logic of natural ecosystemic equilibrium while problematising industrial capitalism as an aberration of nature. Indigenous peoples were reimagined through a new environmental anthropology to deploy concepts of custodianship and environmental management; sophisticated technical–cultural forms of intelligence that kept their primitive economies in balance with their ecologies (Evans-Pritchard 1940; Steward 1955; Rappaport 1967; Vayda & McCay 1975; Harris 1975).

Colin Turnball's *The forest people* (1961) and Jean Liedloff's *Continuum concept* (1975) featured among a wide range of popular ethnographies that praised the Edenic virtues of pre-industrial cultures as a source of redemption for modern cultures, and this has been a persistent feature of environmental thinking ever since. But Roy Ellen's (1986) usefully corrective essay 'What Black Elk left unsaid' showed how such idealisation failed to properly consult the longer anthropological record, and how this demonstrates barely any environmental consciousness or ecological equilibrium among pre-industrial cultures, and if anything, the opposite. It squared with an emerging view from palaeontology that in the late Palaeolithic and early Holocene, hunting and gathering ancestors may have been implicated in megafaunal population extinctions in Eurasia and Australia, and also how indigenous burning across wide sweeps of continental Australia and north America had formative impacts on local environments, certainly not leaving them unchanged (Flannery 1994; Pyne 2001; Franklin 2006a; Prideaux 2012), to say nothing of anthropogenic desertifications and extensive tropical vegetation changes.

However, the linear narrative of a progressive and deepening crisis implicit in most accounts of the Anthropocene ignores human capacities for ecological reflexivity. Just as humanity innovated new forms of exploitation in relation to its own environmental relationship and impact, it also introduced new concepts that drove novel forms of governance and regulation. The most recent phase of the Anthropocene, in the last 50 years, is distinguished by two unprecedented types of change. The first involved unparalleled levels of impact from new technologies, market expansion, unprecedented levels of population growth, settlement, consumption and mobility. The second, a cultural change, involved the arrival of audacious projects (often multiple and contradictory) to take control of entire environments, to order and reorder entire landscapes, on national and international scales, often dedicated to reversing damaging levels of anthropogenic impact and restoring them to the imagined natures of earlier timelines. While these are often seen as recent correctives/adjustments, historians have also discerned more complex, non-linear, reflexive and cultural patterns of association and environmental consciousness, continuities from the early modern period. The sentiment of a lost creation and restorative projects certainly predate the arrival of the scientific concern with biodiversity, for example, with its underlying millenarian promise of provisioning, regulation and cultural service to the planet and humanity (Thomas 1983; Schama 1995; UN Convention on Biological Diversity 2013).

During the later Anthropocene, humanity has arguably entered into new forms of animal domestication on ever-greater scales. Rather than the slower, earlier forms of domestication of individual animal species for pastoralism, farming, hunting and companionability, from around the mid-18th-century audacious national projects aimed to re-order and to control entire wild animal communities too (Ritvo 2012).

Unlike the increasingly generic, corporate, multinational and global organisation of capitalist expansion, the projects to control the nature of wild animal populations were mostly subject to the orderings and governance of smaller territorialisation, those of states or nation-states and were thus culturally and politically more variable. Animals, especially, were subject to processes of colonisation, nationalisms and the subjectivities that formed around them (Macnaghten & Urry 1998), and these produced complex biopolitical contestations emanating from the ebb and flow of identity formation among national assemblages of indigenous and emergent migrant interest groups. Animals found themselves as metonyms and metaphors across a range of ethnic and ethical divisions.

In a country such as Australia, where the present-day dominant vegetation was once an insignificant set of subspecies specialising on margins of disturbance, which then spread and broadened with the advent of natural and anthropogenic fire disturbance (as a new eco-logical norm), the possibility of identifying a community of species as the proper nature of Australia is problematic. This nature cannot be re-garded as a point before anthropocenic disturbance and change but the very result of it (Franklin 2006b). Nature that was once taken as 'pri-mordial' and given the romantic gloss of 'wilderness', was nothing of the kind (Franklin 2006a). Accounts of the sustainable custodianship of Australian landscapes therefore undersell the role of Aboriginal people because they created and managed their landscapes (Pyne 1992). But they also insidiously placed Aboriginal people in nature, rather than culture, thus setting the timeline after first settlement, and hence plac-ing the responsibility for custodial management in the hands of settler society.

Since there are no rules for the enrolment of history into the emer-gent stories of nations, Australia has been able to exercise a highly selective strategy in the adoption of elements from its 19th-century past. The acclimatisation of British and other 'choice species' was organ-ised locally in each of the independent colonies that formed Australia at Federation in 1901, and yet this audacious anthropogenic restructuring of wildlife has also been effectively and retrospectively 'othered', attributed instead to its period as a British colony. In this way, Australian environmental histories claim for nation formation and the gathering pace of Australian nationalism, the very opposite: the championing of native species and the species cleansing of 'invasive' exotics. Although Australian scientists were firm backers of acclimatisation, this is also rarely conceded over the past 20 years or so. Instead they have mounted a counter program of even greater magnitude to bring wild Australia under great 'domestic' control, through protective strategies for 'native' animals that belong and 'invasive exotics' that do not.

Increasingly, modern nation-states everywhere began to modify their modes of relating to wild animals that came within their borders, in part as a result of new subjectivities that responded to new objectifications of the world and its natures. Nations differed in the strategies they followed and in part this was because there was a choice of concepts on offer. In the next section I want to concentrate on two in particular, those of ecosystem and landscape.

## Ecosystem

The ecological and environmental sensibility deriving from the logic and ethics of ecosystem only came into being through the work of early phytogeographers such as Alexander von Humboldt and the context of their work in the progressive re-territorialisation of the planet into nation-states and national/colonial territorial estates (Pyenson & Sheets-Pyenson 1999). Through the unusually broad perspective that this early 19th-century imperative demanded, von Humboldt was able to perceive something that had not come into view from previously laboratory-based and intensive studies of specimens: discrete 'natural environments'. His global classification of vegetation zones became the foundation for new models and language for the modernising natural world.

Such vegetation zones became part of the broad-acre accounting, and modernisation programs of newly emerging nations and their colonial acquisitions, driven to new intensities by the competitive scramble for global positioning and ascendancy (Gellner 1983). Inevitably science was funded at new levels for this work and thus became enrolled into nationalist rationales and logics. Harriet Ritvo (2012) has shown how this tilted species taxonomic revisions and discriminations on nationalistic lines, but it went further than this, to an environmental taxonomy of land itself as ecosystems of varying but always contiguous magnitudes (Bailey 2009).

In 1935 Arthur Tansley was the first to publish the concept of ecosystems though arguably his prime purpose was only to assert the relational nature of species, and not, as they often were, as isolated from the specific contexts (the very constitution) of their systemic exchanges of energy and materials. However, his 1926 book *Aims and methods in the study of vegetation* showed how closely phytologists such as Tansley were tangled up in nation-formation processes: it was edited for the British Empire Vegetation Committee and urged the production of a complete vegetal inventory of the empire for 'efficient management'.

Despite this, it was not until the mid-20th century or so that this abstract modelling of ecosystem was extrapolated and was mapped onto contiguous definable territorial units. This move was not made on scientific grounds, but for national estate management purposes (Lugo et al. 1999; Bailey 2009). While understandable from a practical point of view of conservation, we must beware slippery heuristic concepts sliding from their origins as abstraction to a new life posing as 'reality'. This is where they fall foul of Alfred North Whitehead's (1997, 51) fallacy of misplaced concretism, that argues that simulations or models that help us understand real-life processes should not be confused with the way life really is. This is problematic where one extrapolates from a theoretical understanding of a system of connections among species, materials and elements to a taxonomy of the entire planet and then again to a delineation of territorially bounded spaces through mapping. And yet, this is exactly what happened. Lugo et al. recognise that classifying ecosystems into ecologically standardised and contiguous units is an important step towards efficient ecosystem management, but the potential danger was fixing that which is properly fluid, contingent and abstract, and changing 'systems of association' into 'communities of belonging', building parallel constitutions in nature that mirror and match those of citizenship or 'belonging to territory'. It has the dull stamp of bureaucratic rationality, rather than the lightness, deftness and specificity of the original science it corrupted/parasitised and it is ironic too. As Whitehead wrote some six years before ecosystem appeared:

> That 'all things flow' is the first vague generalization which the un-systematized, barely analysed, intuition of men has produced ... Without doubt, if we are to go back to that ultimate, integral experience, unwarped by the sophistications of theory, that experience whose elucidation is the final aim of philosophy, the flux of things is one ultimate generalization around which we must weave our philosophical system. (Whitehead 1929, 208)

So ecosystem became a new object and ordering for the Anthropocene, an object for the moral refiguring of nature, its governance and its custodians (pan-national governmentality and science) and its 'members'. It provides for the possibility of mapping legitimate membership of nature onto nation; native animals onto citizenship and citizenship onto endemic nature. It provides an objectively based refiguring of territory and its proper management, based on a reality that overarches and overrides the cultural and the 'interested' political claims of individual groups. In actor–network theory terms, ecosystem is a powerful mediator. It is a powerful mediator that covers its tracks.

As a new object, ecosystem created new subjectivities around it (see Pickering 2008). In Australia it created agencies who administer belonging and not-belonging to territory, national organisations, conservation biology strategies and guidelines, and management criteria for national estates (Franklin 2011). It created new natural territories alongside protection, reintroduction and culling policies. It created criteria for 'National Research Priorities' (managing the 'long term use of ecosystemic goods and services' and 'Safeguarding Australia' by 'counteracting the impact of invasive species') and the status of animal as priorities (or pests, depending on the refiguring), hence also new subjectivities and centres for researchers and universities and, in turn, 'privileged research' outputs that fed into aesthetic and environmental assumptions for new environmental subjectivities in civil society (Australian Research Council 2013).

## Landscape

In many ways it was always problematic to fit humanity (and anthropology) into ecosystemic ways of thinking (Folke et al. 1998; Galaz et al. 2007), and so it is not surprising that other objectifications of humans and nature came into being. As Folke et al. (2007:30) put it:

> Few have analyzed the interactions between social systems and key structuring processes in ecosystems. In many volumes on resource management and environmental studies humans have been treated as external to ecosystems. By contrast, studies of institutions have mainly investigated processes within the social system, treating the ecosystem largely as a 'black box.' Analyses of institutions seldom explicitly deal with linkages to functional diversity; key structuring processes, and resilience (capacity to survive disturbance) in ecosystems.

There is very little species consistency in the way humans relate to non-human elements of any given ecosystem, with culture and local reflexivity playing a major role in determining environmental behaviour. Specific groups have been modelled into ecosystemic studies, but they fit awkwardly at best into the notions of both 'system' and 'eco'. Many human groups are not bounded by the ecological content of a given ecosystem domain but range across many or are connected and dependent on many through trade, political patronage and broader social organisations (tribes, nations, empires). In this sense they are rarely merely elements in a given ecosystem, though they can and do have an impact on any one. It is in the nature of ecosystems to return to a state of equilibrium following disturbance (though this fixedness was somewhat increasingly relaxed over the past 30 years to recognise evolutionary change). But because humans are typically changing ecologies permanently and relatively quickly, they are often seen as an externality and posing a risk to their systemic qualities. Part of the problem is the temporal dimensions of ecosystem that has tended to be defined as a 'state of affairs', including biodiversity as measurable at any given time, a snapshot in essence, and to a degree its resilience, its ability to return to that state following disturbance. In fact, it is only an artefact of experimental or empirical time (the time of their measurement) and the extrapolation of those measurements across time, into future time was only ever a modelling exercise. However, because humans are both environmentally reflexive and disturbing, adding and taking away elements constantly, their nature and environmental presence can only be measured through significant periods of time, so it is extremely difficult to assign them to the concept of ecosystemic belonging (Brown et al. 2007). In emic terms, indigenous and other long-settled peoples' histories span periods long enough for temporal presence and for change to feature in how they think about, and relate to, their history and identity. The concept of landscape is just such a term from Europe, in use in Britain from as early as the fifth century among farming cultures, to denote human living spaces, community lands or humanised natures. The Aboriginal term 'Country' references an analogous concept in use among Australian hunter-gatherers, meaning a 'nutritious landscape' in words of Deborah Bird Rose:

> From my studies with Aboriginal people I have developed a definition of country which starts with the idea that country, to use the philosopher's term, is a nourishing terrain. Country is a place that gives and receives life. Not just imagined or represented, it is lived in and lived with. (Rose 1996)

It is not surprising that landscape as a concept arrived at more or less the same time (1934) as Tansley's ecosystem and as a critique of the very idea of natural environments. As the American geographer Preston E James wrote in his now-famous essay on landscape, 'Natural environment, natural setting or natural landscape all suggest a separateness of man from the other parts of the area', and thus 'it seems closer to the truth to think of man as a part of nature – as one element among others which together form the pseudo-organic unity which we call landscape' (James 1934, 78–79).

Subsequently, the concept of landscape as it has been identified in historical ecology was an explicit alternative to ecosystem as an objectification of the world and its natures. According to William Balée (2006), for example, landscape is a central term used in historical ecology to situate human behaviour and agency in the environment.

Whereas ecosystem is from systems ecology, landscape derives from historical geography and cultural ecology. Historical ecology explicitly revised the notion of ecosystem, and replaced it with landscape. While an ecosystem is static and cyclic, a landscape is historical. While the ecosystem concept views the environment as always trying to return to a state of equilibrium, the landscape concept considers transformations to be inevitable processes of evolution. In this way landscapes do not return to a state of equilibrium, but are palimpsests of successive disturbances over time. Stephen Pyne (1992) showed in respect of Aboriginal burning, how humans have been responsible for the emergence of new landscapes, new lived-in ecologies. More broadly, historical ecologists argue that human disturbance has often created greater biodiversity, and to do so was often also an explicit aim (Sax et al. 2002; Brown et al. 2007). Human-made and maintained grasslands are one example (Kirkpatrick 1999).

If ecosystem references wider territorialisation, governance and etic concepts such as biomes, landscape references detailed life worlds of experience, complexity, culture, knowledge and change – including emic concepts such as 'Country'. Here the landscape is a perpetually changing, physical manifestation of history. Whereas ecosystems often suggest zero sum games, where the loss of some elements are understood as gains to (external) others, landscapes can be non-zero sum games where more complexity can be accommodated in win-win solutions. Or not. The work of Sax et al. (2002, 2003) has shown that biodiversity per se is often increasing at local levels and that introductions often exceed extinctions:

> The asymmetry holds even on islands and insular habitats. Within the last few centuries following European colonization, relatively few insular endemic plant species have become extinct, whereas invading species have approximately doubled the size of island floras – from 2,000 to 4,000 on New Zealand; 1,300 to 2,300 on Hawaii; 221 to 421 on Lord Howe Island, Australia; 50 to 111 on Easter Island; and 44 to 80 on Pitcairn Island. (Brown et al. 2007)

While the combined metaphor of ecology-plus-system can slide to imply 'legitimate community' and 'belonging', landscape suggests open, fluid, contested and volatile occupation; it also speaks of the possibility of sharing, accommodation, tolerance, succession and evolution.

Neither ecosystem nor landscapes are 'natural' entities but human concepts that arose in the Anthropocene to help us make sense of changing/changed worlds and guide action. They can be likened, as guides to action, to the concepts of 'nativism' and 'naturalisation' and both of these signify the continuing relevance of nationalism in the lives of animals living wild and national projects to manage them (Peretti 1998; Fetzer 2000; Smith 2011).

In the section below I illustrate these points in a comparative analysis of the place of feral cats in the UK and Australia. Both types of modern societies have introduced cats from outside but their resulting designation, status and treatment is symptomatic of the types of societies they are rather than the behaviour of the animals per se. It shows how ecosystem and landscape have become anthropogenic orderings or strategies for two very different forms of nationalism.

## Nativism and naturalisation: strategies for the Anthropocene and the case of feral cats

On 16 September 2012, I 'googled' the term 'feral cat', first with the addition of 'UK' and then with the addition of 'Australia'. The domestic cat *Felis catus* is a non-native species to both countries but, as an animal designated 'feral' they are treated in opposite ways in the two locations. In the UK case their ferality is a cause for concern, prompting almost universal action for their welfare and care. In Australia it is the opposite: it prompts action for their extermination and control. In Australia the feral cat is largely disliked because of its assumed adverse impact on native species. A similar case might as easily be made in the UK, but is not.

The total UK cat population is in excess of ten million; about one million of those are feral. In Australia the feral population is estimated to be between 14 and 18 million, but given its greater land area, the population of feral cats in the UK is actually substantially denser than it is in Australia (4.1 per square km, compared with between 1.8 to 2.4 square km. Their potential impact on native animals would seem more substantial, but there is practically no scientific effort put into in their eradication and control, nor public support/demand for it, despite substantial public support for all native animal categories at risk. By contrast, in Australia anti-cat sentiments are widespread and substantial, from the political leadership through to most levels of civil society (Smith 1999), including art (for example, Kaye Kessing). Smith (2011) provides evidence that supports the idea that feral cats are widely considered to be 'un-Australian' in civil society.

## Cats and the UK

Cats were introduced as a high-status animal to Britain during the second half of the first millennium BC, part of the zoological remix of Europe, the Middle East and North Africa wrought by the Roman Empire (Davis 1987, 182). Although the cat has lived in Britain for more than 2000 years there are good reasons why it might not be tolerated. Contemporary British culture is focused on the care and welfare of its native animals and there are few other places where birds, in general, have become such objects of aesthetic veneration and national pride. Many species are threatened and although their decline is often the result of anthropogenic disturbance of their habitats, large numbers of feral cats could provoke zealous calls for control, if not extermination.

Native Australian birds feature strongly in the iconography of feral cat risk, yet it is difficult to find any evidence that Australians share the same degree of enthusiasm (or obsession) for birds as the British. In Britain there is one bird species for every 376 square km compared with one for every 6,348 square km in Australia. Two-thirds of all households claim to feed birds and a recent estimate of the numbers of active birdwatchers put it at three million (Unwin 2005). More than one million Britons are now members of the Royal Society for the Protection of Birds, paying a minimum of £36 per year for the privilege. It is the 12th largest charity in the UK.

Small mammals such as shrews, moles, dormice and harvest mice in the UK also evoke great concern and these, too, are potentially threatened by feral cats where their population densities are high. Cats kill more than 44 species of wild bird, 20 species of wild mammal, three species of reptile and three species of amphibian, but anxiety about feral cat welfare is far more significant than anxieties about cat predation on native species (Woods et al. 2003, 176). It is extremely difficult to find evidence for significant levels of condemnation from either science or civil society. According to the Royal Society for the Protection of Birds there is no necessary relationship between high levels of predation and species decline (RSPB 2012). In the image set for feral cats in the UK pulled up by Google there were no inflammatory, dramatic pictures of bird kills.

It is instructive that the first ten hits in a Google search, using the key words 'feral cats UK' showed that welfare for feral cats was the dominant theme (and repeated in the case of the subsequent ten). Eight out of the first ten hits were sites promoting welfare and care for feral cats. Only two did not have welfare as a primary aim. One, the Game and Wildlife Protection League provides advice on controlling feral cats for its landowning members who invite shooters onto their land as paying clients. Visitors to its site are reminded that feral cats are a protected species that enjoy exactly the same welfare legislation provisions as those kept as pets. This is not the case in Australia, rather the opposite.

## Feral cats in Australia

Cats were introduced into Australia shortly after first settlement in 1788 where they accompanied the expansion of farming along the coasts and deep into the interior. Their role was to contain small mammal damage, largely from introduced mice and rats, and many left the confines of farms and settled areas when mouse, rat and rabbit numbers reached plague proportions locally. These spontaneous informal introductions were augmented by planned releases outside of the settled areas to control plagues of mice and rabbits (Dickman 1996, 1).

The first ten items revealed from a Google search using the terms 'feral cats Australia' were all concerned with three main themes: their negative impact on Australia, their control and their eradication. Negative impacts were multiple: environmental (against native animals, biodiversity), social (for example, against Indigenous peoples' cultural values), health (as possible transmissions of disease, such as rabies) and agricultural (in the transmission of diseases affecting livestock). These themes characterised the first 100 items found. The top 20 sites consisted of approximately 50 percent government organisations, 20 percent research organisation sites, 20 percent media sites and ten percent information sites. Further down the list were many voluntary organisations specialising in environmental work, hunting, farming and pest control. In the top 50 sites only three were not negative.

This suggests that the case against feral cats is being made by the most powerful and influential organisations in Australia. It is very difficult to find any other view being expressed and since scientists and scientific organisations are endorsing the government case, they tend to be the ones most cited by the most powerful media organisations that in turn, find their way into educational and other sources of information on the environment.

According to Tim Flannery, only on a few islands and coastal places has the cat made a serious impact on native wildlife, and, at the other extreme, the island of Tasmania has not lost a single native animal due to cats, despite their presence for more than 200 years (Wild Visuals/Discovery 2002). A Tasmanian study concludes that 'there is little evidence that cats in mainland Tasmania are having a significant negative impact on the native fauna' (Schwartz 1995, 59). And for the mainland, as a whole, Flannery argues that 'there is no scientific evidence to say that feral cats were solely responsible for the loss of any native species', and there are studies that demonstrate this very clearly (Jones & Coman 1982, 537–47). But belief in the serious environmental damage done by cats seems completely unimpeded by the paucity of firm scientific evidence for it. It neither impedes scientific inference nor political demands for eradication policies. So, for example in his definitive *Overview of the impacts of feral cats on native fauna*, Chris Dickman argues that:

> acceptable evidence for impact would be any demonstration that cats have caused a decline of 25% or more in the population abundance or geographical distribution of any native species. *Unfortunately, unambiguous evidence of this kind does not exist.* (Dickman 1996, my italics)

It is not clear why the compelling test case of Tasmania does not count as 'unambiguous'. This might equally suggest that feral cats can live in ways that do not threaten wildlife excessively.

It may well be that the cat has done more damage than science has managed to document, but even if that is the case, it is hard to understand why such a strong response is made without more compelling evidence.

All designated serious pests have been declared Key Threatening Processes and this designation, in turn, requires the writing of a threat abatement plan (TAP). A TAP exists for the feral cat, yet the plan states that:

> Convincing evidence that feral cats exert a significant effect on native wildlife on the mainland, or in Tasmania, is scarce . . . There is no evidence of feral cats causing extinctions in mainland Australia or Tasmania. (Australian Government, Department of Environment [State of the Environment] 2006)

A further complication surrounds what Australia can and should do if more compelling evidence were found to support widely held suspicions. The current TAP for the feral cat states that: 'Although total mainland eradication may be the ideal goal of the cat TAP, it is not feasible with current techniques and resources (Australian Government 2008, 5). And according to the Australian Government (2011) 'Even if cats are removed from an area, it is quickly recolonized.'

In recent years scientists have become more concerned with feral cat impact in northern Australia especially, but also in critical places elsewhere. Where dingoes and foxes have increasingly been killed by 1080 poisoning cat populations grow and endanger small mammals. While serious, this does not show that cats cannot live in a sustainable way, merely that whatever balance is achieved is being destabilised by human interventions. It has also been shown that in experiments those native animals are at risk when reintroduced into areas with cats. But this only shows that cats will predate vulnerable populations where they are already strong.

I was also told that in northern Australia before the removal of Aborigines to missions their hunting of the cat would have checked cat populations from rising to levels that would cause decline in native mammal populations.

On governmental and scientific sites and others influenced by them, the feral cat is invariably shown in emotive photos, with a colourful native bird kill, rather than its more typical prey, the introduced mouse, rabbit or rat (Schwartz 1995).

Nick Smith (1999) cites numerous examples of Australian politicians making use of anti-cat sentiment in Australia. A Liberal MP, Richard Evans, made a call in 1996 for the eradication of all feral cats by 2020 and the introduction of native pet species.

Janet Holmes á Court's speech in the 1998 Constitutional convention placed anti-cat emotions alongside the most sentimental of national attachments:

> We need the smell of eucalyptus in this and the feel of red dust. We need to have the feel of swimming in the sea and all those things that make us feel so passionate about this country and love it so much – eating beef and no feral cats. (Smith 2011, 120)

According to Philip Smith and Tim Phillips (2001, 325) 'What has emerged clearly over the past 20 years is growing symbolic potency of the *un-Australian* in the vocabulary of public life in contemporary Australian society'. It is not hard to find the idea of un-Australian being attached to feral animals (see Smith 2011). This could be expressed as that to be 'properly' Australian one 'should' not only want to protect native species but also, as migrants, one should relinquish any sentiments or sympathies for alien animals from other countries of origin. According to Professor Rob Morrison, Chair of the Anti-Rabbit Research Foundation:

> These un-Australian sympathies [for rabbits, cats etc] seem to be rooted in our European background and Americanised culture. But ... native species deserve our support not for arbitrary reasons but for empirical reasons, independent of human benefits. (Morrison 1996)

Nick Smith disagrees:

for many conservationists (and people who would not think of them-selves as such), getting rid of feral biota (and reintroducing native ones) is a way of making the country *and themselves* more Australian. (Smith 2011, 7)

## Nationalism and animals in the UK

Introduced aliens were often appreciated or deplored in the same terms that were applied to human migrants.

Harriet Ritvo 2012, 1

Australia and the UK illustrate how different processes and timescales of nationhood can affect the life chances and social signification of feral cats. The UK's long national history characterised by 'unifications' and 'additions' generated a generally eclectic and cosmopolitan sense of national identity. Introduced and unmonitored wild populations established themselves both before and after initial nation formation in the tenth century under King Athelstan and they were assimilated into the native animal community alongside the long list of migrants, traders and invading humans who brought them (Davis 1987; Wood 2000). The dormouse and the brown hare were introduced during the Romano-British period while the rabbit and fallow deer were intro-duced by the Normans. These and most other animals introduced sub-sequently are all naturalised British. Analogously, in Australia the feral dingo is considered a native animal, but tellingly only because it was es-tablished here before 1788 and was an animal significant to Aborigines.

Later animal introductions came from Britain's colonial territories as trophies and treasures. This largely positive attitude towards the ex-otic continues to influence their designation and treatment, often to the detriment of British nature. The North American grey squirrel is held responsible for the decline of beech woodland in southern England. It is widely considered a nuisance and a pest, and it is (contentiously) blamed for the decline of the red squirrel, but its status is mostly pos-itive and it is protected. As the Young Persons's Trust for the Environ-ment puts it: 'the grey squirrel is a pretty, appealing and entertaining little animal'.

From the late 19th century on the numbers of zoos and private menagerie collections proliferated, and, inevitably, many animals escaped. The Australian red-necked wallaby, Sika deer, grey squirrel, Reeves muntjac deer all formed viable colonies in the early 20th century with very little social comment and certainly not a concerted effort to remove or vilify them. Coypu escaped fur farms in the 1930s and mink followed in the 1950s. Ring-necked parakeets escaped aviaries and established several very popular colonies in the late 19th century. The new colonies have been given adoptive names: the Kingston parakeet, the Twickenham parakeet and the London parakeet.

The history of the social signification of introduced feral animals in the UK offers a way to understand attitudes to feral cats. There is only a very weak differentiation between categories of native and introduced animals; introduced animals that can live as wild animals tend to be freely naturalised and adopted; introduced animals are routinely tolerated and if anything held up as special and attractive; their associations with high status has continued. Critically, introduced animals enjoy the same rights and protection extended to domestic animals. These relate to the UK's relatively early transition into nationhood, its nation formation arising from the unification of opposing elements and its subsequent imperial history of adding yet further elements. Subsequently it has a tendency towards the tolerance of difference, a curiosity for difference and an aptitude for the workable social articulation of difference and a significant tradition of civil liberties and freedom. UK attitudes to feral cats embody these values. Their liberty and freedom, their being at large in the country among native animals does not offend their sense of identity and social solidarity because their standing was more determined by their individual rights as animals. They are thus seen instead as 'at risk' and evoke pity and charity.

## Nationalism and animals in Australia

In Australia, the opposite case can be demonstrated. In 1788 Australia was settled by Britain as a series of colonial extensions of its expanding empire. By the mid-19th century a free settler society was established but their association with Australian nature, particularly its animals, was unsettling. They were odd, aesthetically jarring, lacking utilitarian value and in such abundance that farming was often only marginally viable.

Between 1852 and 1894 every colony (now states) formed prestigious acclimatisation societies comprised of the scientists, the governing class, landowners, and the educated.

In 1853 acclimatisation was still a new French idea, the brainchild of French anatomist Isidore Geofroy Saint-Hillaire:

> The prospect was nothing less than to people our fields, our forests and our rivers with new guests; to increase and vary our food resources, and create other economical or additional products. (quoted in Franklin 2011, 76)

Their subsequent acclimatisations were on an epic scale and momentous (Rolls 1969). From approximately 130 animal species at least 81 animal species have established wild populations (Bomford & O'Brian 1995). As it proceeded alongside other intensifying developments in agriculture, forestry and mining, the scale of change and loss of familiar nature saddened an emerging generation of Australian-born artists, writers and academics who became increasingly vocal and supportive of indigenous animals and their habitats. Assimilation societies gradually lost their appeal.

In 1901 Australia became a nation. The act of nation formation and the creation of single territorial entity created a category of animals, native animals that were united by virtue of their novel inclusion within its boundaries. Biology only reckons territorial belonging from membership of specific ecosystems but, in Australia, a sense of belonging and precedence is given to national belonging (Clout & Veitch 2002).

The case made against the cat, that it does not belong and did not evolve in Australia, is coherent enough. However, many native species could also be considered feral or pest species on the ecological grounds used for cats, yet most are not. The 'science' of feral animals recognises its largely social basis when it admits that 'there are few socially acceptable control techniques to reduce native animal damage (Bomford & Hart 2002, 9).

Equally, some introduced feral animals such as the brown trout and various deer species, are strongly associated with Australian social elites and, despite being very environmentally invasive, as in the case of the trout, and damaging as in the case of the deer, they seem to escape environmental persecution (Franklin 2011). Again, it seems that ecosystem is used inconsistently, a smokescreen for self-evidently social processes.

Tellingly perhaps, among most Aboriginal groups the same antipathy to introduced animals is absent (Rose 1995; Trigger 2008). During much of the colonial and postcolonial period Aboriginal people were effectively displaced from almost all areas of Australia and in fact, were not 'Australians', either before Federation or after as they were not admitted as citizens until 1967. It is logical that the concept of native animals and thus feral animals was meaningless to them. More significant to them was the concept of Country, the nutritious landscape they occupied and worked with. In this sense their cat belonged to Country.


## Conclusion

Like Donna Haraway, maybe Australia is searching for 'a concept of agency that opens up possibilities for figuring relationality in social worlds, where actors fit oddly at best, into previous taxa of the human, the natural or the constructed' (Haraway 1991a, 21). It is perhaps ironic that through their common objectification of the world as landscape people of the UK and the Aborigines have more in common with each other as naturalisers than they do with nativistic settler Australians, though I would not wish to place the UK and Aboriginal practices in the same taxa. Clearly naturalisation can only properly apply to those who hold to concepts of 'nature' and are organised as nation-states. However, Australia seems set to become less wedded to ecosystem and more to landscape. Environmental historians such as Robin and Bashford now write Australian history on the far longer and more contested times of landscape (Bashford 2013; Robin 2013). In his *New nature* Tim Low recognises that the biology of feral animals in their habitat is complex and subject to adaptive processes and that they have been the scapegoat for anthropogenic disturbances and declines in native species. He wrote:

It strikes me that our concepts of nature often do not match what we actually see. Native animals are taken to prefer their natural habitats and natural foods when often that is not true. The words 'nature', 'natural' and 'wilderness' end up misleading rather than informing us about the natural world. (Low 2002, 10)

And this point is very important. The entire conceptual edifice of the Anthropocene holds up a supremacist view of human agency against a largely passive and frail nature whereas we should be taking inspiration from Low, Haraway and others who emphasise the opposite, the potent, unruly and exuberant liveliness of non-humans. As concept the Anthropocene is unhelpfully humanist. And already the scientists I have spoken with know that the cat is here to stay and will become naturalised in time. They also know that their native animals will adapt to them as animals mostly do, though they do worry that some may not be given the time in which to do it. This is where they now see their role, in identifying where complex, non-ecosystemic aberrations disturb further their disturbed landscape and leave some small creatures exposed and requiring protection. This includes many possibilities, from bizarre racial policies that remove Aboriginal hunters whose habitus involved cats as food, to farmers pursuing tight supermarket-driven margins with 1080 bait, to the arbitrary random natural genetic mutations in the mouths of Tasmanian devils and their absence as cat predators.

## Works cited

Australian Government, Department of Environment (State of the Environment) (2006). Indicator: LD-40 Current research into pressures and contributions of naturalised introduced species'. Retrieved on 30 April 2015 from http://www.environment.gov.au/node/22383.

Australian Government (2008). Cat abatement plan 2008. Retrieved on 18 May 2015 from http://www.environment.gov.au/biodiversity/threatened/publications/tap/predation-feral-cats, pp5–9.

Australian Government (2011). The feral cat. Retrieved on 18 May 2015 from http://www.environment.gov.au/biodiversity/invasive-species/publications/factsheet-feral-cat-felis-catus.

Australian Research Council (2013) Discovery projects: national research priorities. Canberra: Australian Research Council.

Balée W (2006). The research program of historical ecology. *Annual Review of Anthropology* 35: 108–39.

Bailey RG (2009). *Ecosystem geography*, 2nd edn. New York: Springer.

Bashford A (2013). The Anthropocene is modern history: reflections on climate and Australian deep time. *Australian Historical Studies* 44(3):329–440.

Bomford M & O'Brian P (1995). Eradication or control for vertebrate pests? *Wildlife Society Bulletin* 23: 249–55.

Bomford M & Hart Q (2002). Non-indigenous vertebrates in Australia. In D Pimental (Ed). *Biological invasions: economic and environmental costs of plant, animal, and microbe species*. Boca Raton, FL: CRC Press.

Brown JH, Sax DF, Simberloff D & Sagoff M (2007). Aliens among us. *Conservation Magazine,* April–June, 8(2). Retrieved on 18 May 2015 from http://conservationmagazine.org/2008/07/aliens-among-us/.

Certini G & Scalenghe R (2014). Is the Anthropocene really worthy of a formal geologic definition? *The Anthropocene Review,* 17 December, doi: 10.1177/2053019614563840.

Clark N (2011). *Inhuman nature*. London: Sage.

Clout MN & Veitch CR (2002). Turning the tide: the eradication of invasive species. In Veitch CR & Clout MN (Eds). *Occasional Paper of the IUCN Species Survival Commission*, No. 27: 1.

Crutzen P & Stoermer EF (2000). The 'Anthropocene'. *Global Change Newsletter* 41:17–18. Retrieved on 29 April 2015 from http://www.igbp.net/download/18.316f18321323470177580001401/NL41.pdf.

Davis SJM (1987). *The archaeology of animals*. London: Batsford.

Dickman CR (1996). Overview of the impacts of feral cats on native fauna. University of Sydney for Australian Nature Conservation Agency, Environment Australia.

Ellen R (1986). What Black Elk left unsaid: on the illusory images of green primitivism. *Anthropology Today* 2(6): 8–12.

Evans-Pritchard EE (1940). *The Nuer*. Oxford, UK: Oxford University Press.

Fetzer JS (2000). Economic self-interest or cultural marginality? Anti-immigration sentiment and nativist political movements in France, Germany and the US. *Journal of Ethnic and Migration Studies* 26(1): 5–23.

Flannery TF (1994). *The future eaters: an ecological history of the Australasian lands and people*. Chatswood, NSW: Reed Books.

Folke C, Pritchard L, Berkes F, Colding J & Svedin U (1998). *The problem of fit between ecosystems and institutions*. IHDP Working Paper No. 2.

International Human Dimensions Program on Global Environmental Change, Bonn, Germany.

Folke C, Pritchard L, Berkes F, Colding J & Svedin U (2007). The problem of fit between ecosystems and institutions: ten years later. *Ecology and Society* 12(1): 30. Retrieved on 24 April 2015 from http://www.ecologyandsociety.org/vol12/iss1/art30/.

Franklin AS (2006a). The [in]humanity of the wilderness photo. *Australian Humanities Review* 38: 1–16.

Franklin AS (2006b). Burning cities: a posthumanist account of Australians and eucalypts. *Environment and Planning D: Society and Space* 24(4): 555–76.

Franklin AS (2011). An improper nature? 'Species cleansing' in Australia. In B Carter & N Charles (Eds). *Human and other animals: critical perspectives* (pp 195–216). Basingstoke, UK & New York: Palgrave Macmillan.

Galaz V, Hahn T, Olsson P, Folke C & Svedin U (2007). The problem of fit between ecosystems and governance systems: insights and emerging challenges. In O Young, LA King & H Schroeder (Eds). *The institutional dimensions of global environmental change: principal findings and future directions* (pp147–82). Boston, MA: MIT Press.

Gellner E (1983). *Nations and nationalism.* Oxford, UK: Blackwell.

Haraway D (1991a). The actors are cyborg, nature is coyote, and the geography is elsewhere: postscript to 'Cyborgs at large'. In C Penley & A Ross (Eds). *Technoculture* (pp21–26). Minneapolis, MN: University of Minnesota Press.

Harris M (1975). *Cows, pigs, wars and witches: the riddles of culture.* London: Hutchinson & Co.

James PE (1934). The terminology of regional description. *Annals of the Association of American Geographers* 24(2): 78–92.

Jones E & Coman BJ (1982). Ecology of the feral cat, *Felis catus (L.),* in south-eastern Australia. *Australian Wildlife Research* 8: 537–47.

Kirkpatrick J (1999). Grassy vegetation and subalpine eucalypt communities. In JB Reid, RS Hill, MJ Brown & MJ Hovenden (Eds). *Vegetation of Tasmania.* Flora of Australia Supplementary Series, No. 8. Canberra: Australian Biological Resources Study.

Liedloff J (1975). *The continuum concept.* London: Duckworth.

Low T (2002). *The new nature.* Camberwell, Vic.: Viking/Penguin.

Lugo AE, Brown SL, Dodson R, Smith TS & Shugart HH (1999). The Holdridge life zones of the conterminous United States in relation to ecosystem mapping. *Journal of Biogeography* 26: 1025–38.

Macnaghten P & Urry J (1998). *Contested natures.* London: Sage.

Morrison R (1996). Ockham's Razor. *ABC Radio National* broadcast, Sunday, 17 November.

Peretti JH (1998). Nativism and nature: rethinking biological invasion. *Environmental Values* 7: 183–92.

Pickering A (2008). *New ontologies*. In A Pickering & K Guzik (Eds). *The mangle in practice: science, society and becoming* (pp8–18). Durham, NC: Duke University Press.

Prideaux G (2012). Australia's megafauna extinctions: cause and effect. Retrieved on 18 May 2015 from http://www.australasianscience.com.au/article/issue-may-2012/australias-megafauna-extinctions-cause-and-effect.html.

Pyenson L & Sheets-Pyenson S (1999). *Servants of nature: a history of scientific institutions, enterprises and sensibilities*. New York: WW Norton and Company.

Pyne SJ (1992). *Burning bush: a fire history of Australia*. Sydney: Allen & Unwin.

Pyne SJ (2001). *Fire: a brief history*. London & New York: University of Washington Press and British Museum.

Rappaport RA (1967). Ritual regulation of environmental relations among a New Guinea People. *Ethnology* 6: 17–30.

Revkin AC (2011). Confronting the Anthropocene. *New York Times*, 11 May. Retrieved on 30 April 2015 from http://dotearth.blogs.nytimes.com/2011/05/11/confronting-the-anthropocene/.

Ritvo H (2012). Going forth and multiplying: animal acclimatization and invasion. *Environmental History* 17: 404–14.

Robin L (2013). Histories for changing times: entering the Anthropocene? *Australian Historical Studies* 44(3): 341–49.

Rolls EC (1969). *They all ran wild*. Sydney: Angus & Robertson.

Rose DB (1995). Land management issues: attitudes and perceptions amongst Aboriginal peoples of Central Australia. Central Land Council Cross Cultural Land Management Project.

Rose DB (1996). *Nourishing terrains: Australian Aboriginal views of landscape and wilderness*. Canberra: Australian Heritage Commission.

RSPB (The Royal Society for the Protection of Birds) (2012). Are cats causing bird declines? Retrieved on 21 April 2015 from http://www.rspb.org.uk/makeahomeforwildlife/advice/gardening/unwantedvisitors/cats/birddeclines.aspx.

Ruddiman WF (2003). The anthropogenic greenhouse era began thousands of years ago. *Climatic Change* 61(3): 261–93.

Sax DF, Gaines SD & Brown JH (2002). Species invasions exceed extinctions on islands worldwide: a comparative study. *American Naturalist* 160: 766–83.

Sax DF & Gaines SD (2003). Species diversity: from global decreases to local increases. *Trends in Ecology and Evolution* 18(11): 561–66.

Schama S (1995). *Landscape and memory*. London: HarperCollins.

Schwartz E (1995). Habitat use in a population of mainland Tasmanian feral cats, *Felis catus*. Grad. Dip. Science Thesis. Hobart: Zoology Department, University of Tasmania.

Smith N (1999). The howl and the pussy: feral cats and wild dogs in the Australian imagination. *Australian Journal of Anthropology* 10(3): 288–305.

Smith N (2011). Blood and soil: nature, native and nation in the Australian imaginary. *Journal of Australian Studies* 35(1): 1–18.

Smith P & Phillips T (2001). Popular understandings of 'UnAustralian': an investigation of the un-national. *Journal of Sociology* 37(4): 323–39.

Steward JH (1955). *Theory of cultural change: the methodology of multilinear evolution*. Urbana, IL: University of Illinois Press.

*Ten million wildcats* (2002). Motion picture. Wild Visuals/Discovery. G Steer & A Ford (Dirs).

Thomas R (1983). *Man and the natural world: changing attitudes in England 1500–1800*. Harmondsworth, UK: Penguin.

Trigger DS (2008). Indigeneity, ferality, and what 'belongs' in the Australian bush: Aboriginal responses to 'introduced' animals and plants in a settler-descendant society. *Journal of the Royal Anthropological Institute (NS)* 14: 628–46.

Turnball C (1961). *The forest people*. New York: Simon and Schuster.

UN Convention on Biological Diversity (2013). Global Platform on Business and Biodiversity. Retrieved on 18 May 2015 from https://www.cbd.int/business/.

Unwin B (2005). Out of hiding: how Britain has become a nation of twitchers. *Independent*, 21 February: 23–24.

Vayda AP & McCay BJ (1975). New directions in ecology and ecological anthropology. *Annual Review of Anthropology* 4: 293–306.

Whitehead AN (1929). Process and reality: an essay in cosmology. Gifford Lectures Delivered in the University of Edinburgh During the Session 1927–1928. Cambridge, UK: Cambridge University Press.

Whitehead AN (1997 [1925]). *Science and the modern world*. New York: Free Press.

Wood M (2000). *In search of England*. Harmondsworth, UK: Penguin.

Woods M, McDonald R & Harris S (2003). Predation of wildlife by domestic cats in Great Britain. *Mammal Review* 33(2): 174–88.

# 5

# The matter of death: posthumous wildlife art in the Anthropocene

*Vanessa Barbay*

Networks of tarmac roads slicing up habitats and wildlife corridors are permanent marks on landscape-bodies that have emerged during the Anthropocene. Roadkill, individual bodies of non-human species that lay by the side of the road, are the silent victims in their thousands, causing roads to seem more like scars on populations of wildlife. Motivated by a desire to engage with the dead animals I encounter (that have been struck by vehicles or bullets) I have questioned what it means to be a painter of animals. Finding the methods of representationalist colonial natural histories and art theory to be complicit with destructive forms of multispecies engagement, I have studied Australian Aboriginal wildlife painting and looked to contemporary theory to develop processes more attentive to non-human agencies. Here I describe the move away from traditions in painting and natural history specimen collection as well as the discovery and experimentation involved in my novel form of artwork production. I also describe how my arts practice is influenced by the enfolding of these global ideas with my personal in-

Barbay V (2015). The matter of death: posthumous wildlife art in the Anthropocene. In Human Animal Research Network Editorial Collective (Eds). *Animals in the Anthropocene: critical perspectives on non-human futures*. Sydney: Sydney University Press.

troduction to taxidermy, as practised by my Hungarian-born father (an amateur naturalist).

Through a long research process, I have developed a form of image making that attempts to realign the agency of the art 'object' with the dead animal subject. As outlined in previous work (Barbay 2013, 52, 56) the production of each work begins when I arrange deceased animals on canvas stretched across bed frames or a trampoline and leave them to decay. The bodies leave a trace or print on the canvas that I then refer to as the 'shroud'. This, I suggest here, enables me to 'collaborate' posthumously with the animal subject, allowing an intensification of material-affective relations between the animal and the artist. This method also serves as an interrogation of the traditional subject-object relationship within representation. Being attentive to the death of animals, and the materiality of a body's disintegration brings into molecular and temporal perspective our shared mortality and fragility, particularly in the context of the Anthropocene themes of extinction and permanent marks on material bodies. As the body falls apart, the multiplicity contained in the complex animal organism is revealed while simultaneously an image emerges containing the dissipated subject.

## Materiality of animal bodies in art practice

My research has revealed that three animal-art practices in Australia: Aboriginal X-ray art, natural history illustration and taxidermy, deal with the materiality of bodies and the indexical qualities of the art or museum object in quite different ways, which have roots in human cultural assumptions about wildlife. In this section I compare these assumptions, also in reference to contemporary art theory relating to non-human species.

As I have discussed elsewhere (Barbay 2013, 53) I spent some time in 2009–2010 living and working in the Kunwinjku-speaking Aboriginal community of Kunbarlanja in Western Arnhem Land. My work focused on artists at Injalak Arts and Crafts and their Ancestors whose art and culture often centres on animals. In his study of Kunwinjku-speakers in Western Arnhem Land, *Seeing the inside* (1996), Luke Taylor explains that:

Kunwinjku control knowledge of the Ancestral world according to a model of what they call 'inside' and 'outside' knowledge. They use the term *kun-yarlang* or the English 'outside' to refer to the most public meaning of things, and the term *mandjamun* or the English 'inside' to identify more restricted knowledge and sometimes secret meanings. (Taylor 1996, 10)

Another level of inside knowledge or *mandjamun* may be interpreted from X-ray paintings. Kunwinjku consider some aspects of particular landscapes to be the body parts of Ancestral beings, who are often in animal form, which means animals in paintings can often represent specific places, their bodies becoming a kind of map (Taylor 1996, 228–33). *Bininj* attribute social agency to paintings of animals considered *djang* (dreaming), as the body parts represented are consubstantial with sacred sites. This symbolic link between the animal subject and country particular to the artist, such as *djang* (dreaming) sites they may be responsible for, also reflects the presence of the particular animal species at the site. The concept of *Umwelt* developed by theoretical biologist Jacob von Uexküll to describe the *'function-circle* of the animal' (Uexküll 1926, 126), is useful here. X-ray painting is a form of representation where human and non-human animal 'function-circles . . . connect up with one another in the most various ways' (Uexküll 1926, 126).

In contrast to the 'connectivity' themes of Aboriginal art, colonial natural history's 'memorialisation' of animal bodies captures what anthropologist Deborah Bird Rose calls 'the arrogance [and violence] of the act of conquest' (Rose 2000, 40). Each stroke of the brush, loaded with pigments from a foreign land, unleashed a different *Umwelt* driven by an imperial dissection and dismemberment of the natural world. As Richard Neville notes regarding natural history, 'it is no coincidence that this movement towards the ordering and description of the natural world should parallel the emergence of England as a dominating power across the globe' (Neville 2012, 11). Natural history painting has become the signifier of human animals who ultimately deny the function-circles of non-human animals. The immediate sentimental agency of natural history art objects implicates them in the development of 'the contemporary factory farm, ubiquitous fast-food outlet, or the steady annihilation of nonhuman species' (Williams 2009, 223), marking them as signs heralding the mass consumption of our animal kin. Natural history's obsession with cataloguing and collecting, particularly in the 19th and early 20th centuries, is made manifest in vast museum and private collections of animal and plant specimens. These are an anthropogenic reordering and plundering of wild species, a cultural trope implicating naturalistic representations of animals (dead animals in particular).

The attitude towards and treatment of animals by my father Tibor are also fundamental to the nature of my art practice and its development (see also Barbay 2013, 56). When he established our family home within my mother's Ancestral lands on Jervis Bay in Yuin Country on the east coast of Australia, he still raised animals for the family to eat and, as was customary in his culture, he also hunted locally to supplement our diet with wild animal meat. I spent a lot of time playing in the animal enclosures mimicking their behaviour becoming accustomed to the smell of animals and the taste of their food. Incongruous with this intuitive behaviour was my father's teachings, which viewed animals as food and involved instruction in the handling of a gun when hunting kangaroos or rabbits and the killing of chickens by decapitation with an axe.

He also taught me how to use the bodies of hunted animals, roadkill and even family pets for creative work in the form of tanning and taxidermy. Initially my father's interest in preserving the natural world centred on butterflies and beetles, which he caught in a net or jar before pinning them into position for display cabinets and later resin paperweights and gear-nobs. He then became known locally as the 'spider-man' due to his collection of spiders. He was particularly obsessed with the most deadly of the species and began breeding them to become expert about their needs and features. The family home still houses this small natural history museum created by my father. In trying to gauge the impressions such events made on my subconscious I have discovered strong feelings of responsibility towards animal bodies that I see in my environment. There is also an associated compulsion to collect them, and a sense of comfort when surrounded by animals or their preserved bodies. Although abhorrent to most animal advocates, taxidermy has a powerful presence due to its strength as an iconic sign infused with indexical validity, although the accuracy of the animal self-captured is obviously questionable, considering the approxima- tions made during the taxidermy process. When Merle Patchett asked,

> When is it that an animal becomes an object? In the case of a taxi- dermy mount, is it at the moment of death? When it is added to a collection? Or mounted? Or put on public display? And is it even useful to think of a taxidermy mount as an object? (Patchett 2006, 6)

She evokes their agency as 'persons'. In previous work I have described taxidermy as 'a chemical process with an unnatural logic that mends the borders of a body already beyond itself' (Barbay 2013, 56). My understanding of the medium as working against nature and my com- pelling childhood misconception of company among the dead led me to deconstruct the idea of a mounted specimen through the process of developing the animal shrouds. My perception of self is bound up with the animals that loom large in my subconscious inseparable from my father, who called me his 'shadow'. As French philosopher Henri Bergson counsels, 'we must seek to discover where, in the operations of memory, the office of the body begins, and where it ends . . . an ever advancing boundary between the future and the past' (Bergson 1912, 85, 88). Specimens held in museums are exactly that, bodies caught be- tween the future and the past.

Rather than working with taxidermy, my distress about the sheer quantity of creatures killed for the purpose of collection or study un- derpins my project to deconstruct the cultural acceptance of preserved animals. As Giovanni Aloi notes in *Art and animals*, Damien Hirst's *The physical impossibility of death in the mind of someone living* 1992 involved the killing of a large tiger shark in order to preserve it in a tank of formaldehyde (Aloi 2012, 1–5); while in 2012 Enrique Gomez De Molina went to prison 'for trafficking in endangered and protected wildlife' (Campbell 2012) in order to produce hybrid taxidermy as art objects. While taxidermy is both an index of a once-living animal sub- ject and an iconic process (whereby the sign and its referent have direct visual relationship), my image-creation experiments with decomposing animals reduce the iconic power of the animal sign, thereby using an indexical process to interrogate the objectifying nature of animal representation, often associated with naturalism and its relative taxidermy. Within the disturbing context of an anthropogenic era I allow the body its natural disintegration and return to the Earth.

In line with Giovanni Aloi's comments about 'the crisis of representation that so greatly pervades most contemporary artistic production ... [and the desire] to access a new representational realm' (Aloi 2012, 7) my research focus on the agency of matter through which the subject can 'speak', echoes the intentions of Arte Povera artists in 20th-century Italy whose focus was to 'question the difference between the *use of* material and *being in* matter' (Miracco & Gale 2005, 20). This approach to the understanding and process of making an art object bears a striking similarity to the mythopoeic links made between local earth pigments and the bodily substances of Ancestral beings (who are animal, human animal or composite creatures) by Western Arnhem Land painters. For Kunwinjku artists, these pigments are significant signs of the continuing presence of their ancient Ancestral cosmology embedded in the landscape to which their families have belonged for thousands of generations.

## Translations from tradition to contemporary practice

Through an interrogation of animal-art practices, I discovered that Aboriginal art practices and assumptions have more to offer contemporary animal artists, compared with colonial representations and memorialisation. The following section describes the translation of traditions in indexical qualities of the object and the agencies of wildlife from Aboriginal X-ray art into my own contemporary 'shroud' art practice.

My PhD research project, 'Becoming animal: exploring iconic and indexical representation', led me to Arnhem Land in 2009 and again in 2010 to live in a community famous for the oldest and most extensive rock-painting galleries featuring animals in the world, Kunbarlanja in the stone country of the West. Kunbarlanja had been the site of extraordinary plunder mid-20th century. More than 50,000 archaeological, ethnographic and natural history 'specimens' were collected during the 1948 American–Australian Scientific Expedition to Arnhem Land with more than 2000 ethnographic artefacts, about 13,500 plants, 460 mammals, 30,000 fish, 850 birds as well as insects and reptiles (May 2010, 8). While human remains interred in rock shelters and collected without permission were repatriated in 2011, following controversy, the vast collection of animal remains are still held at the Smithsonian Museum in Washington.

My research is driven by the 'consubstantiality' enacted through painting with matter of particular significance to the subject. I refer to this relationship between entities using the semiotic term indexical. The sign or representation of animal is 'made from' (Lincoln 1986, 5) the animal subject, and this authenticating process gives agency to the art object (Gell 1998, 7) by enabling a transformation comparable to that noted by anthropologist Nancy Munn among Walbiri and Pitjantjatjara groups in Central and Western Australia where 'a sentient being – takes on or produces a material form, an object, consubstantial with himself ... [constituting] shifts between a subject and an object' (Munn 1970, 141). My obligation to gather deceased animals left where killed and place them in a ritual situation mirrors a funerary ceremony particular to pre-Christian Aboriginal societies such as those experienced by Kunwinjku (see Barbay 2013, 56).

In previous work (Barbay 2013, 53–55) I have discussed my acculturation into Kunwinjku culture, during which I accompanied descendants from the Warddjak clan on a three-day journey to their estate of Maburrinj to collect earth pigments or *delek* from *djang* or dreaming sites whose creation stories describe the corporeal origins of pigments. The most common red *delek* is *gunnodjbe* associated with the menstrual blood of Ancestral women accompanying Yamidj (today manifest as the long horn grasshopper) at the site Gunnodjbedjahdjam (Chaloupka 1999, 83). We collected the staining blood-red pigment in the form of a hard rock on an airstrip at Kudjekbinj outstation. A teenaged boy, Tex Badari, or Nawamud, coaxed it into releasing its voluptuous secrets, after laborious rubbing in a wet medium. The ancient dark red ochre paintings deeply embedded into the rock are often described as the work of *mimih* (thin elongated Ancestral spirits who live in the rock crevices of the escarpment and taught the first people how to live) or, I was told, as the trace of a spirit or creator being when they entered the rock (W Nawirridj, personal communication, 2009). Howard Morphy considers the cross-cultural implications of this belief as 'Abo-

riginal equivalents of the image of Christ on the Turin Shroud ...'
(Morphy 1998, 100). These faded but entrenched apparitions are indeed
evocative of shroud stains, like the bloody, decomposing fluid absorbed
by the canvas in my work, the red pigment penetrates the rock and
endures.

In contrast, the rare white pigment we collected called *delek* is the
first pigment to deteriorate in rock paintings. *Delek* is found at a significant Ngalyod *djang* (Rainbow Serpent dreaming) site called Madjanngalkku. Chaloupka indicates 'the nodules of the white pigment,
huntite, are said to be the faeces of the Rainbow Snake' (Chaloupka
1999, 83) and confirms,

> The third and most prized white pigment comes from Madjarngal-
> gun [sic], along the Gumadeer River near Maburinj, in the Warddjak
> clan's estate. It is scooped 'like flour' from the ground, mixed with
> water and formed into large egg-shaped cakes, which when dry are
> traded across the plateau. (Chaloupka 1999, 83)

The *delek*, which seemed to burst from the muddy river banks in balls
of white, was ceremoniously eaten and smeared over the face and hair
by young Dylan Badari (or Ngalwamud) and her aunty Priscilla Badari
(or Ngalbulanj). I put some on my tongue and as it sizzled and liquefied
I realised *delek* was chalk. As Mircea Eliade noted,

> Modern Man is incapable of experiencing the sacred in his dealings
> with matter; at most he can achieve an aesthetic experience. He is ca-
> pable of knowing matter as a 'natural phenomenon'. But we have only
> to imagine a communion, no longer limited to the Eucharistic ele-
> ments of bread or wine, but extending to every kind of 'substance', in
> order to measure the distance separating a primitive religious expe-
> rience from the modern experience of 'natural phenomena'. (Eliade
> 1962, 143)

When we sank down on the slope by the water and began extracting
this delicate pure substance from the brown sticky mud encrusting it,
Nabulanj confirmed Ngalyod the Rainbow serpent slept beneath the
creek bed in the dry season, emerging in the wet 'to shit *delek* from

its arse' (Badari 2010).[1] I imagined this relationship between paint and faeces emerged from the physical sensations it produces as the paste is enthusiastically smeared on the body, on skin warmed in the tropical heat. It was common knowledge that *delek* was medicinal; good for treating diarrhoea, like the earth of termite mounds. I wondered what snake excrement was actually like and later discovered it has a white component like a birds that was equivalent to urine, a semi-solid wad of impurities. It is expelled alone or with the brown component faeces, distinctly reminiscent of the mud in which this *delek* is embedded.

In exploring *What painting is* from a European perspective, James Elkins reasoned, 'Academic painting had a natural affinity with mud and excrement, because of the common use of brown hues and thick varnishes that yellowed and darkened with age' (Elkins 1999, 69). These natural affinities refer not only to visual similarities, but primal physical sensations. Elkins abandons chemistry and thinks about oil painting using the language of alchemy describing the alchemists' search for '*materia prima* ... the First Substance' (Elkins 1999, 71) from which one could begin the great work. Parallels between the experiences of alchemists when finding their *materia prima* and our experience collecting *delek* were a revelation to me. He notes, 'Another name for the *materia prima* was *terra foetida*, "fetid earth" ' (Elkins 1999, 69–70) and claims:

> The *materia prima* is exquisitely, brilliantly beautiful to the person who can understand it for what it is. In the midst of its rotting pile, it shines at the 'true philosopher' with a secret light. The idea that everything begins in squalor and refuse is an old one ... Alchemists actually dug in swamps, and tried to brew turds and urine. (Elkins 1999, 72)

Once extracted and exposed to the air the young white paste immediately begins to dry, its mud seal is broken causing the spherical body structural disruption. The crumbling matter reveals its transitory nature, its momentary lifespan once integrity is lost. Faecal substances of

---

1   Badari G (Nabulanj), personal communication, 20 July 2010. *Delek* from Madjanngalkku in Maburrinj within the Warddjak clan estate.

perfect consistency moistened in the body dry, crack and crumble once expelled.

The yellow *delek* we collected was a sticky mud called *karlba*. As mentioned in my previous work (Barbay 2013, 55), the *karlba* site featured a small waterfall that filled a deep pool. I dived underwater for the golden *karlba* with the children and blindly felt the sand become clay toward the steep banks. Nawamud left his *karlba* handprint on a distinctive rock above the water. The women said *karlba* was the fat of an Ancestral emu, the Maburrinj area being *ngurrurdu djang kunred* (emu dreaming country). *Ngurrurdu* fat is indeed yellow as is *ngalmangiyi* (freshwater longneck turtle) fat, which is a surprisingly bright yellow and rich in flavour. Paintings on Injalak Hill depicting *ngalmangiyi* largely consist of the fatty yellow mud *karlba* and I noted my perception of this rock art contained a bodily awareness due to my cultural encounters with both pigment and fat. The yellow painting became the fatty edible body of *ngalmangiyi* sizzling in its juices after being dug from the mud of a local swamp. The cultural perception that enables a consubstantial connection to be made between human and non-human animal Ancestors and mineral pigments is indicative of an indexical understanding of the world as a transformative and generative place. According to the anthropologist Bruce Lincoln, this understanding of substances is,

> [also] preserved in the ancient literatures of the various peoples speaking Indo-European languages. The general narrative is that a primordial being is killed and dismembered, and that from that being's body the cosmos or some important aspect of it are created . . . [this] 'creative death' . . . is [used] to establish a set of homologies between bodily parts and corresponding parts of the cosmos . . . there is thus posited a fundamental consubstantiality, whereby the one entity may be created out of the material substance of the other. The two are understood as *alloforms*, alternative shapes of one another. Viewed thus, flesh and earth, to take one example, are seen as consisting of the same material stuff. (Lincoln 1986, 2–5, italics in original)

Sacred paintings adorning European churches represent the wounded and crucified Christ, whose body and blood consubstantial with sacraments offered in the form of bread and wine are ritually consumed and are believed to bring everlasting life, facilitating worship conducted through the pages of sacred texts sung to life in hymn. This 'process whereby matter was recurrently transubstantiated from a microcosmic form to a macrocosmic form and thence back again' (Lincoln 1986, 40) substantiates religious and by extension social law connected to an ancient European worldview. Like early Christian icons generated within monasteries as illuminations in holy manuscripts or painted panels and mosaics covering church interiors to facilitate worship, some X-ray paintings of *djang* (dreaming) animals connected with sacred sites are produced 'in more restricted contexts. An important aspect of paintings used in these restricted realms is the transformative Ancestral power of the painting itself' (Taylor 1996, 20), lending a powerful social agency to the painting object.

To unpack the concept of agency in relation to both the animal subject and the art object, I refer to Alfred Gell's anthropological theory of art extrapolated in *Art and agency* that is defined as 'social relations in the vicinity of objects mediating social agency' (Gell 1998, 7). Gell specifies that

> Agency is attributable to those persons (and things) who/which are seen as initiating causal sequences of a particular type, that is, events caused by acts of mind or will or intention, rather than the mere concatenation of physical events. (Gell 1998, 16)

As a merging of subject with art object I find it useful to consider Gell's concept of 'abducted' agency residing in the shroud, the dead matter being the subject that asserts a physical presence as an individual with whom I participate in the studio and who I present as an agent in the world. While decomposition is a physical event and cannot in itself be described as having 'agency', the actions of the animal leading to death, and the particularities of their physical form combine with my encountering their body and initiating 'causal sequences'. In my work I argue that to reinvest agency in a deceased subject through the production of an art object expresses veneration and posits the object as a subject among subjects.

This broadening of consciousness allowing the spectator to 'consider art objects as persons' (Gell 1998, 9) aligns with a return to the plane of immanence, a continual awareness associated with animal consciousness as described by Georges Bataille in *Theory of religion*. This reverses the objectifying conditions imposed by modern industrial society and its excessive production/consumption of objects and materials. Modernity and the military order's empire-building are understood by Bataille to be actively perpetuating 'the positing of the object: the tool' (Bataille 1992, 28). He locates the change from wild to modern in an exponential growth within the realm of object and as Kim Levin notes, art reflected this change, 'modernist art insisted increasingly on being an object in a world of objects' (Levin 1985, 3). Modernity created conditions in which the living can be objectified or, as Bataille notes, 'men situated on the same plane where the things appeared elements that were and nonetheless remained continuous with the world such as animals, plants, other men, and finally, the subject determining itself' (Bataille 1992, 31).

In contemporary human–animal discourse animals are considered subjects rather than objects in art and literature through reflexive terms such as 'agency' and topics such as 'animal histories' – which Erica Fudge reminds us 'is in reality the history of human attitudes toward animals' (Fudge 2002, 6). If representing animals in art is understood in terms of Gell's anthropological theory, artists can be viewed as taking responsibility, as mediators, for returning agency to the subject through a relational cross-species negotiation. According to Gell, the particularities of this transformative negotiation within the process of making and of presenting art enables the art object or 'material 'index' ... [to permit] a particular cognitive operation which I identify as *the abduction of agency*' (Gell 1998, 13). As an indexical sign the shroud is a natural sign of the animal subject due to the '*causal inference* [while] ... Abduction is a case of synthetic inference where ... the *index is itself seen as the outcome, and/or the instrument of, social agency*' (Gell 1998, 13–15, italics in original). The shroud stain thus vacillates in a zone between causal and synthetic inference as evidence becoming social agent.

In representation it is the agency of the artist, the art object being evidence of the artist's 'hand', that holds precedence. In order for the animal self to counteract, challenge or at least be on par with the artist's assertion of self and intentions about their animal subject, an indexical relationship between the animal and the representation (whether mythopoeic or scientific) needs to be apparent. This fine line within representation has been the focus of my work in harnessing the decomposing animal body to produce the indexical animal image. Indexical representation can be found in the work of various artists focusing on animal subjects such as the renowned contemporary UK painting and drawing duo Suzi and Olly who allow predatory animals in their natural habitat to interact with the art object in process. As Ron Broglio notes regarding their work, 'the artist's paper spreads out as surface between the animal, and the artists create a contact zone at the edge of the human and animal worlds' (Broglio 2011, xxx). These indexical traces left by an individual living animal's body and the implications of self and environment then contained in the artwork parallel the indexical traces the dead animal, and the environment in which it decays, leaves in my work. This direct transference of an animal's substance speaks of the animal more than the artist. As Broglio asserts:

> The surface can be a site of productive engagement with the world of animals ... Artists are keenly aware of the optical and physical surfaces that function as the material for making art ... This expressive language formed from the double fold of these surfaces creates conditions for thinking the problem of contact between the 'surface' animal world and our own ... This is a corporeal thinking that risks itself, mind and body, in the acts of encounters that differ with each animal. (The event is not simply singular but a swarm or a pack that multiplies with each engagement.) ... the sites where the human and animal worlds bump against each other, jarring and jamming our anticipated cultural codes for animals and offering us something different. (Broglio 2011, xvii–xviii)

The woven cloth medium through which the decomposing animal has been carried or transmitted in my work captures the resultant contagion in the sense explored by Gilles Deleuze and Félix Guattari. In their essay 'Becoming-intense, becoming-animal' the authors state:

Contagion, epidemic, involves terms that are entirely heterogeneous: for example, a human being, an animal, and a bacterium . . . These combinations are neither genetic nor structural; they are interkingdoms, unnatural participations. That is the only way Nature operates – against itself . . . These multiplicities with heterogeneous terms, co-functioning by contagion, enter certain assemblages: it is there that human beings effect their becomings-animal . . . dark assemblages, which stir what is deepest within us. (Deleuze & Guattari 1987, 242)

The shroud is an assemblage whose becoming exists at every moment as collaboration between human, animal and bacterium, an interkingdom of contagion within Nature's contextual precondition. In this sense, I propose that the shroud becomes a form of representation that uses contagion to validate itself. The canvas as the inert carrier in this repulsive exchange gains authenticity and 'enters into alliance to become-animal' (Deleuze & Guattari 1987, 244). It is the hordes of bacteria that infest the weave that produce a foul odour. Human contact with a corpse is taboo in many cultures once decay is imminent and bacteria multiply.

In collecting pigment from the rather dainty decay of birds, a sense of the diabolical is not evoked. In 2010 I placed a rosella on canvas to decompose. While one of numerous bird shrouds I produced, the result of this particular experiment was startling in several ways. The representation is very clearly a bird and although this may not be obvious to the spectator, the substance, shape and proportions of the bird image evident are particular to the rosella. I found the small colourfully feathered body of this rosella by the roadside in Canberra where many birds and animals can regularly be found dead or dying. In addition to the shroud impression, the bed springs, its frame and the wire mesh holding the body firmly in place also make impressions on the canvas. The site also becomes embedded into the work as the wind blows over the body, adding marks made with dust, leaves and sunlight or by collecting additional matter such as droppings and tree sap. This is how the image of *Rosella* manifested over time. The body ravaged by postmortem processes and exposed to the elements leaves its trace. Time will tell whether these pigments of decay are indelible stains or evanescent hues.

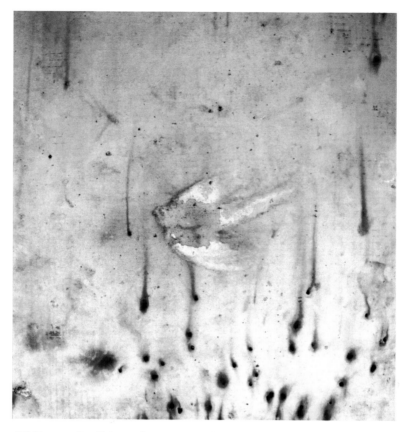

*Gift (Autumn Rosella)* 2010, by Vanessa Barbay. Rosella, sap and rabbit skin glue on canvas; 79 x 82 cm.

The canvas is altered, although more stable than a corpse, it undergoes subtle transmutations, its integrity compromised with the stress and strain it bears in contact with the decomposition process. In this particular case, the tiny organs and minimal flesh and fluid oozing into the weave from such a small skeletal cavity remains contained within the limits of its form. This factor for the majority of bird shrouds

I produced adds to their ability to sublimate their abject constituents. In her exploration of cultural and social order Mary Douglas notes,

> Any structure of ideas is vulnerable at its margins. We should expect the orifices of the body to symbolise its specially vulnerable points. Matter issuing from them is marginal stuff of the most obvious kind. Spittle, blood, milk, urine, faeces or tears by simply issuing forth have traversed the boundary of the body. So also have body parings, skin, nail, hair clippings and sweat. The mistake is to treat bodily margins in isolation from all other margins. (Douglas 1966, 121)

The clearly defined boundaries of a recognisable silhouette will enhance the iconic capacity of the figure prompting an understanding of the sign as a representation.

Contrary to the rosella's involuntary contamination of the cloth, adjacent areas of purity remained where the canvas was sheltered from the rain and dust under water-repellent feathers. Wings and tail void of putrefying tissue being of an inert and enduring substance, continued to protect and to nurture after the death and decay of their generative avian body. It had been a wet start to autumn in 2010, with wind and rain gusts depositing dust around the rosella's perimeter. This soft shadowy edge extended its reach quite miraculously beneath the body defining the wing against the torso. The eucalyptus tree bled its sap generously, evenly distributing honeyed droplets whose pigment spread into the fabric creating smoke-like stains. The tree had never before provided such an offering to the shrouds it had witnessed over the years, and has not marked its presence so overtly since. The exposed canvas has darkened a shade and collected the subtle gridlines of the mesh, which appears to rust or collect dirt in order to print itself unevenly over the surface. Another bird has left its dropping too, as though signing the work 'made by a bird', for how can I say that I represented this bird? It is more accurate to say that, at my invitation, the rosella represented itself.

The cultural power of the figure as an iconic sign is indisputable, but matter is also loaded with iconic and indexical meaning. The role of the figure in painting is not limited to its historical or cultural associations. My greatest challenge during this project was the one posed by painting itself, which, as Deleuze notes concerning the paintings of Frances Bacon, is 'to extract the figure from the figurative' (Deleuze 2003, 8). My leap in this regard was to facilitate the agency of the dead animal subject, enabling it the privileged role in making the first marks as the figure 'represented'. Deleuze claims 'It is the confrontation of the figure and the field, their solitary wrestling in a *shallow depth*, that rips the painting away from all narrative but also from all symbolisation' (Deleuze 2003, xiv). What remains is matter as pure sensation.

## Works cited

Aloi G (2012). *Art and animals*. London & New York: IB Tauris.

Barbay V (2013). Becoming animal: matter as indexical sign in representation. *Antennae, Painting Animals II* 26 (Autumn): 52–58.

Bataille G (1992). *Theory of religion*. R Hurley (Trans). New York: Zone Books.

Bergson H (1912). *Matter and memory*. NM Paul & WS Palmer (Trans). London: G Allen & Co Ltd; New York: Macmillan.

Broglio R (2011). *Surface encounters: thinking with animals and art*. Minneapolis, MN: University of Minnesota Press.

Campbell J (2012). Artist Enrique Gomez De Molina sentenced to prison for wildlife smuggling. *Huffpost Miami*, 2 March. Retrieved on 30 April 2015 from http://www.huffingtonpost.com/2012/03/02/ artist-enrique-gomez-de-molina-sentenced-wildlife_n_1316807.html?

Chaloupka G (1999). *Journey in time: the 50,000-year story of the Australian Aboriginal rock art of Arnhem Land*. Sydney: Reed New Holland.

Deleuze G (2003). *Francis Bacon: the logic of sensation*, 3rd edn. DW Smith (Trans). London & New York: Continuum.

Deleuze G & Guattari F (1987). *A thousand plateaus: capitalism and schizophrenia*. Minneapolis, MN: University of Minnesota Press.

Douglas M (1966). *Purity and danger: an analysis of the concepts of pollution and taboo*. London & New York: Routledge.

Eliade M (1962). *The forge and the crucible: the origins and structures of alchemy*, 2nd edn. S Corrin (Trans). Chicago, IL: University of Chicago Press.

Elkins J (1999). *What painting is: how to think about painting using the language of alchemy*. New York: Routledge.

Fudge E (2002). A left-handed blow: writing the history of animals. In N Rothfels (Ed). *Representing animals*. Vol. 26 (pp2–18). Bloomington, IN: Indiana University Press.

Gell A (1998). *Art and agency: an anthropological theory.* Oxford, UK: Clarendon Press.

Levin K (1985). Farewell to Modernism. In R Hertz (Ed). *Theories of contemporary art.* Englewood Cliffs, NJ: Prentice-Hall.

Lincoln B (1986). *Myth, cosmos, and society: Indo-European themes of creation and destruction.* Cambridge, MA & London: Harvard University Press.

May SK (2010). *Collecting cultures: myth, politics, and collaboration in the 1948 Arnhem Land expedition.* Lanham, MD: Altamira Press.

Miracco R & Gale M (2005). *Beyond painting: Burri, Fontana, Manzoni.* London: Tate Publishing.

Morphy H (1998). *Aboriginal art.* London: Phaidon Press.

Munn ND (1970). The transformation of subjects into objects in Walbiri and Pitjantjatjara myth. In RM Berndt (Ed). *Australian Aboriginal anthropology: modern studies in the social anthropology of the Australian Aborigines* (pp141–62). Nedlands, WA: University of Western Australia.

Neville R (2012). *Mr JW Lewin: painter and naturalist.* Sydney: NewSouth Publishing.

Patchett MM (2006). Animal as object: taxidermy and the charting of afterlives. Retrieved on 30 April 2015 from http://www.blueantelope.info/files/pdfs/AnimalAsObject2.pdf.

Rose DB (2000). The power of place. In KA Neale (Ed). *The Oxford companion to Aboriginal art and culture.* South Melbourne, Vic.: Oxford University Press.

Taylor L (1996). *Seeing the inside: bark paintings in Western Arnhem Land.* Oxford: Clarendon Press.

Uexküll, J von (1926). *Theoretical biology.* DL Mackinnon (Trans). London: Kegan Paul, Trench, Trubner and Co Ltd.

Williams, Linda (2009). Modernity and the other body: the human contract with mute animality. In K Kitsi-Mitakou, Z Detsi-Diamanti & E Yiannopoulou (Eds). *The future of flesh: a cultural survey of the body.* New York: Palgrave Macmillan.

# 6

# A game of horseshoes for the Anthropocene: the matter of externalities of cruelty to the horseracing industry

*Madeleine Boyd*

Seven hundred horses a month – many young fillies and colts bred for racing – are slaughtered at two Australian abattoirs and shipped overseas for human consumption, including to Europe, the centre of the horsemeat scandal. The majority are slaughtered in Queensland at Caboolture's Meramist Abattoir, where 500 horses are processed each month. A further 200 a month are killed at a South Australian abattoir, Samex Peterborough (formerly Metro Velda). Thousands more are processed at 33 knackeries across Australia for pet meat and hides each year, with industry reports indicating the annual cull totals about 40,000 (Thompson 2013).

> Observe, when the starting barriers are flung back, how the race-horses in the eagerness of their strength cannot break away as suddenly as their hearts desire. For the whole supply of matter must be mobilized throughout every member of the body: only then, when it is mustered in a continuous array, can it respond to the prompting of the heart. (Lucretius [c 99–55 BC], cited in Kane nd, 10)

Boyd M (2015). A game of horseshoes for the Anthropocene: the matter of externalities of cruelty to the horseracing industry. In Human Animal Research Network Editorial Collective (Eds). *Animals in the Anthropocene: critical perspectives on non-human futures.* Sydney: Sydney University Press.

## A brief paleoethology of multiple-bodied entities

About 2 to 5 million years ago *Equus caballas* emerged alongside other equine variations, eventually fading into extinction (Davis 2007). It was 2.5 million years ago that humanoids emerged, developing late (compared with *Equus*) into behaviourally modern *Homo sapiens* approximately 70,000 years ago.[1] Various meetings across the plains between *Homo* sp. and *Equus* sp. occurred across the millennia as co-habitants of the landscape or predator and prey, until a closer relationship began. About 6000 to 4000 years ago, the horse and human dyad (McGreevy et al. 2009) emerged as a multi-bodied entity[2] in what is now southern Russia (Davis 2007). This entity went on to enact narratives of epic proportions, of which stories were told to later generations: a Chinese emperor with Night Shining White, Alexander the Great with Bucephalus, Caligula with Incitatus, and so on. These dyadic memories are heralds of civilisation, and foundations for the age of Anthropocene. Now riding astride the entire planet, Anthropos cast aside equid companions like old shoes. Approximately 200 years ago the horse–human dyad was wrenched apart, and these multiple entities began to die out over large parts of the globe, like *Eohippus* and Neanderthal thousands of years before them. The Iron Horse steam engine became the new beast of burden for the age, and the human–technology entanglement engraved tracks of steel and tar where earth and stone once were. Anthropos continues on in the present day, but not entirely alone. The livestock that engendered the prospering of Humanid forebears have not been cast aside, but still live among civilisations, even if it is at the margins beyond everyday sight. Even as this period of the Anthropocene marches into folds of geological time by making marks on the Earth's surface, our fellow earthlings (Dibley 2012, after Bruno Latour) and citizens of shared history and culture, are making marks and are

---

1   Retrieved on 27 April 2015 from http://en.wikipedia.org/wiki/ Timeline_of_human_evolution.
2   Natasha Fijn's 2011 etho-ethnology within Mongolia, the most ancient, well-known and vast extant horse–human culture, strongly demonstrates the entangled intra-actions across species, bodies, landscapes and culture.

being marked. We are all together caught up in the material–discursive grand narrative of the Anthropocene.

In his discussion of the Anthropocene, Ben Dibley (2012) writes about the supposed economic conditions of freedom, the guise under which freedom has been sought. He remarks upon the externalities of these economics, never really externalities, merely a growing debt. In the industrialised Anthropocene the horse–human dyad at the core of civilisation has been replaced by an economics of greed, with externalities of cruelty in the form of widespread horserace gambling industries. To suppose that 'to be human' is a stable concept is critiqued by discussions within the Anthropocene forum and related posthumanities. It might also be said that what makes us 'human' are our relationships, our existence in a shared manifold of intra-actions (Barad 2007) with humans and other species. As we move away from the horse–human dyad, and more broadly direct relationships of care with domestic livestock, it follows then that what it is to be 'human' is changing; and that a growing debt of cruelty is the externality to the evolution of this life form without a name, unless that name is *Homo destructus*. No species is an island, unless it is an island only. What follows in this chapter is a present-day experiential and research-based account of encounters with the externalities of the horseracing industry in Australia. The process became enfolded with development of an artwork for the Australian Animal Studies Group (AASG) Conference exhibition in 2013. As an artist working under the pressures of the Anthropocene, these encounters concern questions of horse–human entanglements, deconstruction of conundrums in empathy, and a posthuman (new) materialist perspective; a fitting perspective for this mutational, geological, entangled Anthropocene age.

## Bodies killed with kindness

### Research diary entry 1

Turning out of the gates from the New South Wales country estate where I was learning for the first time about natural horsemanship, I turned left instead of right towards the highway and home. Was this curiosity, a will to explore, or destiny? First there was the dead wallaby. Swollen and distended in the evening's granular light. I snapped some arty shots, trying to juxtapose the entity of death against the potentiality of the road surface. Sick, really. Further on just down the hill and around a slight bend, my eyes widened like pancakes. There was a small field over full with 20 or 40 horses. First impression: not much grass, mostly dirt actually, and the horses were just standing around, scattered in small clumps. I slowed the car down, pulled along side the fence and tried to get some quick shots. The property certainly didn't have a welcome sign out, nor was there any open hostility evident. I just felt scared spending too long inspecting and photographing the site. Who knew what the situation was. I stayed long enough to ascertain that the horses looked healthy enough. One or two were munching on the remnants of a lucerne bale. A big pinto, showing no bones, was a bit muddy, but that's par for the course with a white horse on dirt.

The next day, back at my course, I waited for a suitable moment and asked about that property. I knew there must be a little sensitivity to it, because it stood in glaring contrast to these immaculate horse training grounds, and the combined impressions of Zen, wealth and perfection in the unity between horse and human. The short answer was given, 'That's Kim Hollingsworth's property'. Shock! Horror! A Scoop! This name rang a bell as that of one of the leading characters in the well-known Australian true crime drama show *Underbelly* set in Kings Cross, Sydney (our local den of iniquity).

Kim Hollingsworth is the most seductive of characters to begin a narrative with. In real life Hollingsworth was a prostitute who, through the course of the television series, turns her life around to become a policewoman. An attractive actress dramatised and sexualised her role in the television series, lending the 'real' woman an aura of mystique. My mind reeled with interesting possibilities, drawing links between Hollingsworth's life story and the field filled with too many horses. Online investigation revealed that horses had all been 'rescued' by her from horse sales (where 'the doggers' buy horses for pet meat) and other poor situations. So the apparent case of animal hoarding that I witnessed might have revealed the conflicted face of animal rescue. Further research led me straight to controversy, which Hollingsworth seemed to attract like a moth to flame. She was not made popular by her kind gestures towards equines. Instead, the local community clearly vilified her with threats of all manner of hate acts. Repeated calls for the RSPCA[3] to seize her horses and insults of the worst kind appeared in long and scathing written posts from several parties on social media sites at the time I was conducting research (November 2011 onwards).

Emergent equine–human cultural phenomena seem to be 'lifting up', 'connecting with' and 'healing'. Many reports have been published about the ways in which connecting with a horse helps people to navigate difficult 'embodied emotive' states (for example, Bachi 2012; Equine Psychotherapy Australia;[4] various writings of Temple Grandin). From the esoteric spirituality taught at the Epona ranch (Kohanov 2007), to the theatrics of *The black stallion* and *War horse* films and the growing professional area of equine assisted therapies, horses have established a reputation for healing and even 'saving'. I have experienced the overwhelming sensations of seeking and co-healing in the process of encountering and bringing back to health my rescue horse Picasso.

New materialist philosopher Jane Bennett (2011) considers that hoarding is due to the attraction of materials. Hoarders are out-of-the-ordinary people who have a greater sensitivity to the vitality of objects and materials than most. Yet Bennett suggests these people should be encountered with interest and compassion, for their abilities with matter have much to teach us. Meanwhile the horses live for now, but only a marginal existence, a product of multiple pressures, and the source of multiple harms.

## Situatedness of horse rescue in Australia

Zooming out of this narrative, the case study can be contextualised within the naturalcultural situatedness (after Haraway 2008) of horses

---

3   Holland (2013). Royal Society for the Protection of Animals. The organisation in Australia granted with the power to cease livestock and companion animals and prosecute for criminal neglect.

4   Retrieved on 27 April 2015 from http://www.equinepsychotherapy.net.au/.

in Australia. There exist several organisations focusing on horse welfare and horse sports. 'Heavy Horse Heaven' was recently featured in the weekend *Daily Telegraph* (August 2013), and focuses on the draught-type of horse. In this Anthropocene age of the machine, the need for draught horses that used to work pulling heavy loads is generally recreational or for breed preservation. These huge horses fall by the wayside when cost of upkeep outstrips their interest factors or owner/carers fall on hard times, such as drought. Some of the most active and informational groups are the protest group 'Ban Jumps Racing'[5] and 'Coalition for the Protection of Race Horses'[6] in Victoria. Ban Jumps Racing regularly protest and make (illegal) documentary observations of the jumps races that are still legal in this country, much as the horrific National Hunt[7] courses are in the UK. Even beyond the usual dangers of horseracing, jumps racing requires a cavalier attitude towards the injury and death of horses for human recreation and gain. This seems to be out of step with commonly held animal welfare standards, yet the industry continues. A recent media report suggested that more than 40,000 horses each year are sent to the knackeries and abattoirs (Thompson 2013). The report was set in the context of lowering demand from the European horsemeat market after wrongly labelled burger meat was found to be tainted with horsemeat in 2013. Although horsemeat consumption by humans is common in Europe, citizens appeared to have been put off by unexpected horsemeat in their meals (VIZZATA 2012). However the more significantly off-putting information contained in Thompson's (2013) report is the number of horses going to the knackery in Australia. They are mostly rejects from the racing industry: too slow, too many, and too many to rescue.

*Research diary entry 2*

When I rescued Picasso, the most beautiful palomino Arabian stallion many people will ever set eyes upon, he was the defunct remnants of a breeding program that had gone by the wayside of a

---

5  Retrieved on 28 April 2015 from http://banjumpsracing.com/.
6  Retrieved on 28 April 2015 http://www.horseracingkills.com/.
7  Retrieved on 28 April 2015 http://www.britishhorseracing.com/.

woman's life after several 'unwanted pregnancies' (pers. comm.). I was not looking for a horse like that, nor did I know much about horses at all, but when I came upon him I said there and then 'you're coming with me, buddy'. I guess it was lucky for both of us that this was an isolated incident. Since then I have been to the (in)famous Camden Horse Sales just south of Sydney, and I have witnessed the small huddle of horses awaiting their judgement with no trial. $1.30 a kilo, off to market they go. I could not choose which one to save, and I certainly cannot afford to help them all. So instead I left with a heavy heart, documentary images and more pieces of the puzzle of horse rescue. It was on my second excursion to the sales I felt that I had become an 'ineffectual martyr'. Emotionally wounded, materially entangled, but without sufficient materials to alter the outcome of the day.

## Material resonances of racehorses and the performativity of art

It was about the time of events described in Research diary entry 2 and when I was also learning about wastage (standard practice of killing of unwanted racehorses) and *nanny mare foals* (unwanted foals bred to induce milk production in mares for racehorse foals) that my farrier (horse hoof trimmer) turned up one day with a bucket overflowing with horseshoes. I knew that particular farrier service serviced the local racecourses, which is why they always kept me waiting an hour or two. The shoes came in many different sizes, fitted to individual horses, and most were worn down on the edges where the hard but vulnerable hoof walls would meet the track or bitumen. Bits of manure and hay still clung to the metal, with a faint whiff of urine. Not an altogether unpleasant combination for the horse enthusiast. For me, these objects rang with potency. Here was the sedimentation of so much experience, a collision of histories and entities: horses, farriers, jockeys, hoofs pounding, the bookies, the turf, the stable, the speed, the pomp and celebration, the suffering. Taking these resonant objects to my art studio, I contemplated the horseshoes for some time. I wondered how I could transform these sedimentations to do new work in which they could exert the power of their inherent agency. Traditional modes of sculptural form, like welding or arranging did not seem to allow the fluidity required. An active incarnation would onlysuffice. I started to think with the agential realism of Karen Barad (2007) and recent developments in the related field of new materialism (for example, Bolt 2013; Dolphijn & van der Tuin 2012).

Within a new materialist conceptual framework, the artist can be conceived as material-philosopher. Theirs is a performative engagement with matter to produce material-discursivity (Barad 2007) in action and form. As an artist with an interest in engaging new materialism and the non-human, I consider objects and actions that emerge from relations between multiple actors to be aesthetic, in other words to become art. I also work with the idea of engaging artworks as sedimentations of multi-agential worlding. Bolt (2013, 4) suggests that the 'artistic relationship embodies a materialist dialectic' in her introduction to new materialism in art theory (although I would replace the dualistic concept of dialectic with a multiple agential term such as Karen Barad's intra-action). Among artists, Cézanne is cited is a forerunner of materialism in his use of paint and the way that he transmitted information from the landscape to the canvas. Wallace (2005, 65) discusses the influence of artist Cézanne's materialist painting on his cultural contemporary DH Lawrence, who said 'Cézanne reacquaints us with matter, while science, reveling in words, facts and figures, denies it'. Tuma (2002, 60) digs deeper still into Cézanne's materialist influences to discover that learning of the early atomistic theories of classical Greek philosopher Lucretius most influenced his work:

> Lucretius wrests a material world with phenomenal qualities, qualities that take place in space and time, and that are there given to vision as nature's forms . . . This is the Lucretius I think might have meant something to Cézanne.

Materialism in art has progressed in waves since Cézanne first made an impression on the academies with his methods and atomistic thinking as he produced multiple paintings of Mont Sainte-Victoire. Karen Barad's theories are based in the updated atom studies of quantum field theory, and are having profound reverberations in contemporary philosophy and art, just as Lucretius did for Cézanne. The point is made here that the agencies of the materials becoming performative in a work of art are of primary aesthetic importance to the (new) materialist artist.

The interplay of my own embodied agencies, with that embodied by the horseshoes, as well as the unsettling phenomena emergent in cultures of horseracing caused me to contemplate the idea of horseracing as a blood sport. The race-day outfits and celebratory occasions popular in the Australian culture of 'going to the races' stand in bizarre contrast with the bloody behind-the-scenes wastage of horses. This sport is nothing new as going to the races is one of humanity's oldest entertainments. In ancient Rome the Circus, as the racetrack was known, was the biggest and finest of buildings. It stood in honour to the powerful contribution of horses to Roman and contemporary culture. The best mounted armies paired with the fastest, hardiest and bravest breeds of horses lent any side the winning advantage in the games of empire (Hempfling [2001] 2011). Chariot drivers of Rome were celebrities and the mad Caesar Caligula elected his horse Incitatus to the senate. This account is from selective Western histories as grand narratives, but does well to illustrate that horseracing is an institution central to civilisations built by the horse–human dyad and helped pave the road towards the Anthropocene. In the world of today, the horse–human dyad has faded away in everyday life, but the races continue.

I have it from a presentation by horse welfare academic Professor Paul McGreevy at the AASG Conference (2013) that Australian racehorses are particularly strong, partly due to the practice of heavy culling, another complication on the road to justice. If we consider breed conservation for a companion species with which our own Western human history is so closely entwined, we must then question where does the balance of ethical breeding lies. I have had a sickening thought about the uncontrolled overbreeding of humans. A subject dealt with, and with adequate curiosity and distaste, in the science fiction film *Gattaca* (1997). Jeff Wallace (2005) writes that Charles Darwin's evolutionary theories and the rise of science led to open discussions of eugenics among philosophers of his time. Horses are not subject to the same ethical codes applied to humans in the 20th and 21st centuries, and so are bred as a genetic project towards a perfection we humans can never apply with candor to ourselves. Bio-artist George Gessert produces interesting commentary on breed development with his living artworks consisting of orchid-breeding experiments.[8] Similarly, in the horse breeding world ever more extreme versions of Arabians, miniature horses, and other breeds are produced. Sometimes these creatures are so exquisite it is hard to believe they are real. My observations of the horse-breeding industry lead me to believe there is little critical inquiry into these practices by the breeders, or society at large. When self-critique does occur it tends to be met with hostility and claims of attack (the institutions of interest protect themselves).[9]

### Affective moments towards interrogating interspecies violence

Coming back to the racing industry, the gears of interspecies war and jealousy are self-evident. Breeders strive for the stronger, faster horse, and covet the stud or mare that produces champions with the most consistency. At the time of writing (September 2013) a highly prized Australian stud stallion in the Hunter Valley, New South Wales, has recently succumbed to colic (twisting of the intestine), 'a tragic death' was reported, a loss of million (of dollars). All the other 'didn't run' horses will not attract a pretty penny down at the knackery, so there would be no loss reported for them. Still the questions hang potently: why? Why do the horses need to go faster? Why do new horses have to be introduced to the track every year? Why are horses pumped with such high-energy feeds and medication that they frequently succumb to colic? Why do we hang the pride of a nation on a fast horse, the fastest horse, but all the others can go to hell? Authors such as Melanie Joy[10] have dealt with the

---

8   George Gessert entry on the Viewing Space website. Retrieved on 28 April 2015 from http://www.viewingspace.com/genetics_culture/ pages_genetics_culture/gc_w02/gc_w02_gessert.htm.
9   Although, it should be noted that during October 2013 on miniature horse-breeding social media sites two lively discussions around breeding and welfare brewed: the excessive numbers of foals and horses on the market during the current economic downturn and debates on the fashionable practice of shaving horses' heads for show presentation. The discussions revealed strong factions, and some vitriolic language, but also good intentions by some breeders and enthusiasts.

factors at play in the whys of inconsistency in our actions towards other species. Yet the grandness of the horse–human connection through history and the urges of beauty and power push this particular conundrum into the realms of poetry and art. Will the dirty laundry of the legends of yesteryear be aired? Napoleon was painted valorously up on his favourite white steed. In truth he was a terrible horseman, and there was more than one 'favorite white steed'. They were replaceable as they fell upon the battle fields all heavenly and pure in pools of red blood and stark against the grey-green mud as they lay tangled or writhing in pain, still wonderful, crashed to the Earth we must now walk upon since we have fallen from their backs. It is useful to pathologise this behaviour? How can we capture the sentiment of such experiences, to show up the dangers of glamour? Could this be the work for art to do?

## Resolution in playful material-discursivity

In a moment of connection between layers of association, or even epiphany, in my studio process, I realised that the racehorse shoes I had gathered from the farrier could be activated through a game of horseshoes. We can take the horses out of the sport and play a game of horseshoes instead. This game facilitates a cross-species experience of skill tests. Human versus human, we ply our skills at the game: and the loser ... will be killed (of course not). The game of horseshoes, along with the game of war with horses, has a long history. Where humans find themselves gathered in groups with little to do, entertainment and contests of skill and betting inevitably emerge. With little else on the battlefield but horses and men, the game of horseshoes emerged in pre-Roman times,[11] and had resurgence in the USA revolutionary war.[12]

---

10   Melanie Joy (2010) suggests in her theory of carnism (opposing vegetarianism or veganism) that conflicting attitudes towards animals might be due to the invisible but powerful network of beliefs and assumptions, coupled with a culture of uncritical acceptance of the status quo, as well as concerted efforts by the industry to hide unsavoury aspects of livestock production and to promote biased health information motivated by profit-making.
11   History of horseshoe pitching. Retrieved on 21 April 2015 from http://www.horseshoepitching.com/gameinfo/history.html.

The game is very simple. It involves tossing horseshoes at a pole sunk into the ground at a slightly forward angle. The closer the shoe lands to the pole, the better the score. There is a wonderful scene in a Robert Redford film (title long forgotten) in which he redeems himself after a criminal life within a group of 'honest cowboys' by displaying great skill and winning a game of horseshoes (Robert Redford also directed and starred in the film *The Horse Whisperer*, 1998). In the film a pivotal moment of greatness, forgiveness and excitement is evoked with a simple game of horseshoes. Today horseshoes is a regulation sport in the USA, with tournaments, prizes and standardised rules. For the AASG Conference I produced a prototype for an interactive game of horseshoes. The game was set up for gallery visitors to play, with a horseshoe pit and the horseshoes collected from racehorses. A video projected onto the backboard of the pit displayed messages and images. The images alternated between those of horses on their way to the knackery, which I photographed at the Camden Horse Sales, and publicity images for flashy racehorses and race carnival events. The projected messages also varied, based on each person's throw result (or that would be the desired effect in a fully operational game): 'hit the backboard: too slow, off to the sales'; 'near shot: start 5 more races'; '3 in a row: out to pasture, become a stud'; 'perfect shot: you're lucky, rescued by a teenager for pony club'.

In considering the possible value of such a game, research shows that video games have a relationship with empathy levels in the player (Belman & Flanagan 2010).[13] Empathy is emerging as a key motive for good doing in the world, attributable to evolutionary, neurological and relational factors (for example, de Waal 2005/6; Gallese 2003). Yet 'games' do not need to be all for fun or profit, they can instead challenge the player in complex emotional ways. Most popular video games show gore and violence. For example, *Resident evil* was promoted in London during October 2013 with a well-rendered and gruesome human butcher shop installation (Crawley 2015). This is a gratuitous entertainment value-driven industry. Whereas developers of more 'gritty' games

---

12   Horseshoe history. Retrieved on 21 April 2015 from
http://www.sportsknowhow.com/horseshoes/history/horseshoes-history.shtml.
13   See www.gamesforchange.org/.

*Game of horseshoes for the Anthropocene* (2013), by Madeleine Boyd. Racehorse shoes, regulation sized horse shoe pit, mini projector, sand and mixed media; dimensions variable. Draft version for an interactive game on the uncertainty of life for the Australian racehorse.

that deal with real-world narratives, suggest that their programs can be conceived as more fulfilling and purposeful, even if they are harder to digest:

> Why would anyone want to put themselves through this? 'For the same reason you'd want to read a novel about something really heavy,' says Ms. Armbruster, a 20-year-old college student. 'There's something really satisfying about experiencing narratives that are outside your own experience.' (Dougherty 2013)

Similarly, a game showing images of down-and-out horses on death row alongside horses that have the best of homes and lots of love, could give a fuller, more intriguing narrative for the player to engage with, compared with the run-of-the-mill *My horse*-type,[14] idealised horse-ownership games, and the ethical emptiness of war games. The contemporary art context allows for this type of experimentation and engagement by curious humans.

The materiality of interacting with horseshoes and provocative images stands alongside literature, traditional studio art forms and activist information campaigns. Literature from Anna Sewell's *Black Beauty* to JM Coetzee's *The lives of animals* do provide powerful explorations of animal life worlds and 'other worlding', evoking imaginative possibilities. From the traditional art world, the power of image is well known. Yet posthumanist theorist Cary Wolfe (2009) makes the point that visual studio arts such as painting and photography are highly anthropocentric. Additionally, art popularised within anthropocentric economies is heavily influenced by the aesthetics of visual marketing. Art for the Anthropocene can be considered to engage other species, work with matter in material-discursive practices, and performatively entangle relations in confronting pressures of concern such as interspecies justice. So I suggest here that physically engaged artworks that draw the observer/participant in by engaging the materiality of the body are actively doing work on many levels of 'other worlding'.[15] Material interactivity additionally fits well with thinking about the horse–human bond of physicality on which, for example, equine assisted therapy is based. The importance of physical or kinetic learning is discussed within 'learning styles' theory (for example, Advanogy.com 2013, after Howard Gardner) and integrated learning theorists (for example, Price & Rogers 2004). Artworks can provide a powerful sedimentation of experience that can be quickly absorbed by the 'viewer/intra-actor'. Once an artwork has been experienced the material-discursive concepts moves performatively into other worlding. There is only

---

14   See www.naturalmotion.com/my-horse/8/.
15   In *When species meet* (2008), Donna Haraway begins by describing 'autremondialisation' as an alternative to globalisation, here anglicised as 'other worlding'.

the need for people to engage for the artwork to do its work for inter-species justice.

## Materiality of blood sports

In developing the potential and motivations of this 'game of horseshoes for the Anthropocene' further, I now consider related literature on the Australian racing industry, thinking around blood sports, and the externalities of horseracing. Phil McManus and Daniel Montoya (2012) present one of the few cultural investigations of the thoroughbred racing industry in Australia, which they also suggest is surprisingly under-theorised. They seek to develop a geography of jumps racing in Victoria through its framing in newspaper media. The two main groups shown to be in conflict are the jumps race industry (including the local town that relies on the races economically) and the activists who assert the races are cruel because frequently horses die on the track. Within this study horses are not passively theorised, as their potential for agencies of varying types is discussed. However the analysis suggests that in the media, industry and activist dialogues horses tend to be anthropomorphised, objectified or essentialised (McManus & Montoya 2012). This is considered somewhat unavoidable as horses cannot speak for themselves. It is suggested here that other ways of representing the horse voice might be a type of cradle-to-grave agency and wellbeing analysis or a framework based in response-ability (after Barad 2007).

An important factor that could not be addressed within the scope of McManus and Montoya's (2012) necessarily limited jumps racing study is that many people in the wider public are not aware of the extent of externality issues around horseracing. As well, if they are made aware by media reports, we can assume there rarely exists sufficient empathic engagement to motivate action to improve the lot of horses. The wider public are neither activists nor working in the horse industry that represents the small sector of horse–human dyad types in an age of Anthropos. So, the task remains for empathic outreach beyond activist internet sites and newspapers. Here I also question the stated neutral position in the McManus and Montoya (2012) paper towards the jumps racing industry. While presenting a balanced view of the human participants in the debate, they fail to accentuate the objective situation that the overall benefit of the racing industry remains in favour of humans and not in favour of horses. There is no balanced perspective if we consider the horses. The key point made in McManus and Montoya's (2012) paper is that people in the industry who 'love' the horses assert that they are saving them from the abattoir. However, this goes back to the externality of over-breeding. Quite possibly these horse lovers actually think of the horses as wastage and lucky to be alive. Having been saved, these horses therefore have a reduced inherent life value (they were going to die anyway). Consider that in this post-horse–human dyad Anthropocene age there is no need for racing. If racing must continue as an industry and an entertainment, proceedings could be slow and safe, with horses of all ages included. Over-breeding and wastage slaughter clearly should be made illegal based on the imbalances of the impact of the industry on horses compared with that on humans. The drunken punters dressed in their Sunday best, or the sad addicts down at the midweek TAB might hardly notice that the horses were 12 years old, and running at 80 percent speed. The spectacle, pomp and ceremony continue. Market-wise, the stifling of supply (breeding) could push up value. Here I am being somewhat cavalier in suggesting that there would be ways to preserve some jobs around, caring for and working with horses, and so preserving the horse–human dyad without need for the torture en masse of horses. I consider a balanced perspective is therefore to grant racehorses cradle-to-grave response-ability.

There is evidence of a push for greater consideration of animal welfare in meat-production industries of Australia. Live export of cattle has received high-profile media attention as have campaigns for with free-range chickens and pigs. For the 'ethical carnist', more products and local businesses based on the free-range standards have opened shop (for example, in Sydney, Australia, Feather and Bone). Perhaps the widespread interest in ethical meat relates to the materiality of ingesting those animals, and hence an inherent bodily empathy. I eat ethical, therefore I am ethical. By contrast the punter at the TAB or the racetrack does not materially connect with the big picture of horseracing. So they do not question the stream of new entrants to the bookies' form and wonder where last season's horses have gone. They do not sense the externalities of cruelty.

During my brief stint working in the racing industry as a stable-hand, for the love of horses, I was young and unaware. I saw only the immediacy of the horses. I was not involved in the breeding, the medicating, the killing, or the syndicates. I just saw 'horse'. Huge, muscular, wild horses that made me feel both awe and frightened for my life. There was unforgettable charm in the predawn parade of horses and jockeys down to the track. The snorting, steaming breath of the great beasts and their hoof beats, the jingle of tack, the heady smell that emanates from a horse, the sense of secret goings on at this place out of step with time in the middle of a 21st-century international city. Yet those were bracketed moments, the moments of glory that sustain the romance. Much the same, race-day goers have their bracketed experience and syndicates, groups of people with money to invest but little horse sense, think the death of their horse is due to some isolated incident of illness or unavoidable accident at the track. They do not know how to question the modes and methods applied to the horse.

## Agencies of the underworld

If the punters were aware and could care, another push away from justice comes from the underworld. Corruption in the horseracing industry has been identified in the media and by industry watchdogs recently and as far back as the famous Australian racehorse Phar Lap's possible poisoning by American gangsters. Underworld money and influence should not be unexpected in a gambling-based industry; an industry designed to reap the rewards of others' miseries, whether they be pensioners at the local TAB or racehorses destined for a short and arduous life. Evidence for undue influence of the violent and 'money-talks' kind, is naturally hard to gather, although a 2008 report suggests it is rampant (Playthegame.org 2012). One clue is given in the reporting of the late Les Samba (who was gunned down in an execution-style killing in Melbourne in 2011), whose daughter 'Victoria – the official face of the Melbourne Cup carnival in 2005 – was the former head of marketing for NSW Racing and was married to jockey Danny Nicolic.'[16] It is not who you are, but who you know.

Taking an evolutionary perspective, Torill Christine Lindstrøm (2010) explores a sensorial opening into the conundrums of blood sports in the arenas of ancient Rome. Analogies can be drawn between the spectacle of the gruesome staged hunts and horseracing carnivals. Differences exist in today's displacement of the spilling of blood more often away from the site of the event instead in front of the audience. Although McManus and Montoya (2012, 409) note that for jumps racing, death can be part of the thrill but 'for other journalists death is treated as part of the entertainment experience'. They cite a journalist from *Herald Sun:*[17]

> Death at the racetrack, as a public spectacle, is different to death at the meatworks or at some far-flung gymkhana. There is no other mainstream modern sport or activity in which death lurks so regularly, where the family can plonk itself on the picnic rug and be thrilled, then horrified.

Lindstrøm seeks to understand the seeming conflict between the attitudes of the Romans to animals in society compared with in the arena, just as many animal activists seek to expose the conflicts in such things as horseracing and wastage, pet ownership and barbecue culture. The strongest theories include that of a culture coming out of the wild. Subsistence and hunting culture were still extant in Roman times. This suggests a co-evolutionary relationship between human and non-human, by which humans could appreciate good qualities in Other and still retain the ability to kill or to alter for survival, hence a survival mechanism of bipolar attitudes and actions. Lindstrøm (2010, 325) refers to the brain science supporting this:

---

16   Drugs, FBI, Mcgurk: racehorse owner Les Samba's mafia connections. *The Sunday Telegraph*, 17 April 2011. Retrieved on 21 April 2015 from http://www.dailytelegraph.com.au/
racehorse-owner-les-sambas-mafia-connection/story-e6freuy9-1226040244161.
17   Stewart, M. Image wins the day in soft new world, *Herald Sun*, Melbourne, 30 November 2009: 65.

> To perceive the connections between pain-blood-death (the PBD complex) during hunting or cruel acts releases dopamine in the brain, creating feelings of joy and pleasure, and general arousal ... Through identification ... with the *venatores* [staged hunts] the spectators could have gone through the emotions of fear, coldness and joy/pleasure in rapid succession. As will be elaborated below, such rapid changes in emotional states have particular effects on human susceptibility ...

Another strong line of reasoning presented by Lindstrøm (2010, 317) is that of bystander effect:

> Part of group and mass experiences is a feeling of reduced personal responsibility, called 'bystander apathy' or 'bystander effect', and is explained as an experienced diffusion of responsibility ... A consistent finding is that the more onlookers are present, the less probable it is that an assaulted individual will be helped. Many *ludi* [arena killing events] had thousands of spectators. This number itself would render it unlikely that anybody would react to defend any animal or human being attacked in the arena.

Have we 21st-century humans moved beyond the gruesome spectatorship of Rome, so primal and contradictory, or are we really not so different, and still acting as these same primal humans, our tendencies thinly veiled by social contracts, or are we different, but still emerging? In *Capitalism: a love story* (2009), filmmaker Michael Moore iterates a point gestured at by Lindstrøm (2010) that ancient Rome was built on slavery, and democracy faded away into the dictatorship of the Caesar over time. The violent spectacular events were put on to placate the poor and the enslaved. The appreciation of violence may have also been deeply entangled with the culture of violent suppression, embodied by those groups and events. Moore suggests that this practice has reemerged in recent decades with Wall Street 'dictator' interests now having a strong presence in the White House. Their agenda is to remove the middle class and re-establish a lower class, ruled by the elite. Tactics of expensive and spectacular political campaigns, using the public media of today, bring to mind the patterns of ancient Rome. Similarly horseracing as an elite sport offered as a get-rich-quick gambling opportunity and appealing to the thrill of violence, works into the kind of elitist-violence-pathos field of cultural emergences suggested by Moore and Lindstrøm.

While the carnist theories of Melanie Joy and the lessons from the Roman spectacle of staged hunts in the arena do provide some reasoning and backstory for the ability of humans to condone violence towards animals, Kristin Armstrong Oma (2010) considers a more developed cultural theory, beyond the raw evolutionary traits of the hunter: that of a social contract between humans and domesticated animals. Janzen (2011) also writes of the ancient contract between humans and domesticated livestock, while interrogating the Earth's ability to sustain humanity's spiralling appetite for animal products.[18] Horses fall into this theoretical approach on a sliding scale of companion animals and livestock, depending on their use variously as transport, pets, entertainment or meat. The ancient social contract, in turn, suggests a duty of care on behalf of the humans, in the case of horses, has led to the global distribution and breed diversity of *Equus caballas* as the European human/equine entanglement expanded to conquer new worlds. This social contract inherently suggests that abuse of a horse would be outside of the contract, particularly if the horse is holding up its side of the bargain. So here we find the inalienable basis for improvement of treatment of horses in the horseracing industry. Given the ancient antecedents of the social contract between human and horse, we can consider that it trumps the lessons of barbarous human evolutionary traits from the arenas. This is particularly true because horses have been

---

18   One last benefit of livestock, perhaps least understood and quantified, is the subtle but powerful attachment of people to the animals – the almost mystical bond of the 'ancient contract'. Rollin (2008), for example, reflected on how 'ranch people often sit up all night for days with a marginal calf, warming the animal by the stove in the kitchen, and implicitly valuing their sleep at pennies per hour!' Cummins (2003), pondering the rewards of looking after livestock, wrote: 'But when I start thinking about how our animals and crops and fields and woods and gardens sort of all fit together, then I get that good feeling inside . . .' Such examples, which presumably occur in countless ways worldwide, imply that humans and societies derive benefits from animals beyond mere monetary value (Janzen 2011, 788).

domesticated and cannot be separated easily from their naturalcultural embeddedness in civilisation. This line of reasoning applies to the frequently raised question 'should thoroughbred breeders have inalienable rights to do as they wish with horses?' The answer appears to be no, and so from this point onward, we act from a position of justice, based in the co-creation of naturecultures between horse and human. This sits well within Karen Barad's agential realism from which we understand that individual agency does not exist a priori, but only through intra-action and mutual constitution with other agencies.

Armstrong Oma (2010, 180) goes further in her analysis as well to describe the horse–human entanglement within the social contract such as the relationship of horse and rider:

> This seamless relationship needs to be entrained. Game (2001, 3; see also Hall 1983) explains this as 'learning to come in tune with', and it demands openness and receptiveness of the other (ie the animal). Humans and horses entrain riding together in a horse–human rhythm, in which they learn how to tune into one another. In this process the human and the horse are in tune together, the relationship is what matters and species are forgotten. 'What horses and riders entrain with is the relation, the rhythm between, the transporting flow, the riding' (Game 2001: 5). Riding well demands that one forgets the human separate self, as riding is 'absorbing horse, taking horse into our body' (Game 2001: 9). When one learns and embodies a motion, the motion is inhabited (Bachelard 1969), thus, through rhythm; horse and rider come to inhabit riding. Consequently, thinking of riding in relational terms moves the focus from the rider as carrying out an action to an understanding of the rider and horse as simultaneously carrying and being carried by each other. Fundamental to this particular form of intra-action is mutual trust.

## Social contracts for post-industry equines

In McManus and Montoya (2012) and Armstrong Oma (2010) strong gestures are made towards theorising the naturalcultural intra-actions of horse and human in horse industries. An extension of theory that this current chapter suggests is the inclusion of the post-industry equines that have seemingly moved out of the social contract per se into the no-man's-land of wastage and rescue. Here we find loose ends, the dirty laundry and the externalities of cruelty. Horse 'use value' has shifted from the heady context of the Thoroughbred Magic Millions yearling sales, to the dubious and dusty realm of the saleyards and neglected back paddocks. They are no longer part of the growth economy. To the consumer, rather like last year's Christmas chocolates in the supermarket, marked down, tempting, but probably rotten. From the human perspective we find in the post-industry arena the rescuers and rehabilitators as well as, according to McManus & Montoya (2012), the jumps racing industry. Leaving the last category out, as they are considered still within the industry of horseracing, in 'rescuers' we find a different kind of human, and different motivational factors. There is no doubt that some professional horse trainers have found an economic niche by retraining high-bred racers for dressage and eventing, to their merit. Yet let us focus on the constitutive outsiders, the externalities, the 'other than expert trainer' rescuers of OTT (off-the-track) standardbreds and thoroughbreds, and the broken-down ex-racers that are well beyond the 'ready-to-retrain' stage.[19] Here we find past winners now forgotten, skinny broodmares full with foal, nanny mare foals discarded too young through the sales, and on the other side humans with high levels of empathy with equines, a strong desire to do what they can and understanding of the gross injustices they are witness to.

What needs to be stated here is that violence is being inflicted upon these rescuer humans by the horseracing industry. Here is the extension from the cavalier attitude of the breeders and 'front-end' industry members who are reaping the monetary rewards, to the 'back-end' battlers who struggle to pay their feed and vet bills for the rescue horses, eventually realising with heartbreaking sincerity poured out through social media sites that they cannot save them all. The romanticism of

---

19  'An assessment of the foot condition [of horses at the Queensland horse abattoir] showed that many of the hooves were overgrown and required farrier attention. An analysis of six hoof indicators, which are generally associated with overgrown and untrimmed hooves, revealed that 80.5% of the horses presented with one or more indicator.' (Doughty 2008, 2)

*Black Beauty* who finds his boy again crumbles into financial stresses, regular calls for donation, and expanding financial membership requirements. Considering that the evidence points to the vast majority of rescues coming from the racing industry,[20] why does the burden of paying for their injustices towards our societal contract with horses fall to members of society at large? The racing industry that gives with one hand in the form of entertainment, jobs and economic growth (as reported by media cited in McManus & Montoya 2012) takes with the other hand economically and emotionally. Undoubtedly there are stress-related health impacts on humans, and those on the horses themselves. So this story comes full circle to make the point that empathic tendencies towards other species become prostituted by greed, as the figures of justice's underworldy reap the rewards of desire: desires to become with other species, as spectator, as fellow animal, as steward.

## Positive experiences-that-matter

To end on a note of possibilities, and as a motivation for change, a relatively new, but very active, horse rescue organisation was the recipient of a community grant offered by a banking institution. The award was based on public votes, and won just ahead of a cancer cause. Heartening to know, because although cancer is an important charitable cause from a humanist perspective, a group of us humans have shown ourselves to be able to think and to vote outside of our own species' direct interests. Likewise, jumps racing has been banned for several years in New South Wales and Queensland (McManus & Montoya 2012), although the battle continues in Victoria. From my work in the studio, I found the confluence of this gathering of knowledge, and my own empathic entanglement with horses sedimented in the material pro-

---

20   My own field observations at the sales and observations following several horse rescue organisations through their online news; and: 'Observations of the types of brands present indicated that 52.9% of the horses processed had originated from the racing industry with 40.0% of the sample group carrying a Thoroughbred brand and 12.9% carrying a Standardbred brand. The remainder of the group (47.1%) had no visible brand.' Doughty (2008, 2).

duction of the game 'A game of horseshoes for the ineffectual martyr'. In the gallery, I felt the artwork was doing good work, even by provoking engaged conversation around the subject of the work. Many of the gallery-goers were willing to try their skill at the game, feeling variously that they would be terrible at it or quite confident, all sending the light aluminum metal objects hurtling through the air to settle around the post or smash against the backboard. Like horses, our human skill at the game varies for many reasons, be they motivational or physical. One woman who came in the see the exhibition at her daughter's behest was taken aback by the story told by the game. It had never occurred to her before that dark secrets lurk in the pleasures of gambling, and that we humans should care for the plight that we inflict upon horses. So I count this as a small but significant success in the creation of critical conversations and experiences-that-matter in the contemporary art gallery.

Returning to the inquiries that motivated this essay, it is clear that engaging with externalities of cruelty that plague the ancient social contract between humans and horses and particularly the present horseracing industry requires confrontation of a complex manifold: a system of intra-actions-that-matter between significant bodies; material limitations;[21] the antecedents of interspecies empathy; and the dangerous allures of fast money and beauty. While this type of analysis may seem fanciful to some, consider the alternative, suggested by the Learn Liberty education website:

> Prof. Sean Mullholland at Stonehill College addresses a classic example of a negative externality, pollution, and describes three possible solutions for the problem: taxation, government regulation, and property rights. The first two options are difficult to monitor and may create perverse incentives. A better solution to overcome the externality is property rights, as described by Ronald Coase. As long as property rights are well-defined, divisible, and defendable, par-

---

21   At the time of writing, 12 March 2014, a prominent horse rescue group in Australia has just announced closure of two rescue facilities due to the current extended drought. This has both reduced available feed in paddocks and caused a rise in the price of commercial feed.

ties can negotiate to reduce the impact of the pollution. (Mulholland 2011)

This kind of anthropocentric, top-down, heavy-handed tendencies of modern-day democracies fail to address the complexities presented by the underworld (organised crime), the will to participate beyond monetary or property motivations (why do audiences continue to attend the races?), and the need for justice-orientated conversations and experiences-that-matter across temporal-cultural-economic situatedness (for example narrative histories of horses and humans). Looking forward, also into the realms of possibility, indications are that the horse–human dyad may return. As the pressures of the Anthropocene deepen and fuel scarcity looms, already in economically marginal areas a rise of horse power is indicated (Church 2014). There is no doubt that the horse–human dyad remains relevant. As evolution continues within the Anthropocene, the only option for survival as we know it is to evolve 'in relation with', and to embrace becoming *Homo sapiens relationata* while allowing *Homo destructus* to fade into the lost temporalities of speciation branches that did not adapt with relations-that-matter and, importantly, in consideration of the externalities of both material and justice kinds to relations that support existence.

## Works cited

Advanogy.com (2013). Discover your learning styles – graphically! Retrieved on 30 April 2015 from http://www.learning-styles-online.com/.
Armstrong Oma K (2010). Between trust and domination: social contracts between humans and animals. *World Archaeology* 42(2): 175–87.
Bachelard G (1969). *The poetics of space*. Boston, MA: Beacon Press.
Bachi K (2012). Equine-facilitated psychotherapy: the gap between practice and knowledge. *Society and Animal* 20(4): 364–80.
Barad KM (2007). *Meeting the universe halfway: quantum physics and the entanglement of matter and meaning*. Durham, NC: Duke University Press.
Belman J & Flanagan M (2010). Designing games to foster empathy. *Cognitive Technology* 14(2): 11–21.

Bennett J (2011). Artistry and agency in a world of vibrant matter. The New School and Vera List Center for Art and Politics. Youtube. Retrieved on 30 April 2015 from http://youtu.be/q607Ni23QjA.

Bolt B (2013). Introduction: towards a 'new materialism' through the arts. In Barrett E & Bolt B (Eds). *Carnal knowledge: towards a 'new materialism' through the arts* (pp1–14). New York: IB Tauris & Co Ltd.

*Capitalism: a love story* (2010). M Moore (Dir). Beverly Hills, CA: Anchor Bay Entertainment.

Church SL (2014). Beasts of burden: targeting disease in Africa's working donkeys and horses. *The Horse*. Retrieved on 12 May 2015 from http://www.thehorse.com/features/34557/beasts-of-burden-africas-working-horses-and-donkeys.

Crawley D (2012). Inside Resident Evil 6's human butchery gallery. *VB Gamesbeat*, 1 October. Retrieved on 21 April 2015 from http://venturebeat.com/2012/10/01/inside-resident-evil-6s-human-butchery-gallery/.

Cummins T (2003). *Feed my sheep*. Bloomington, IN: authorHOUSE.

Davis B (2007). Timeline of the development of the horse. *Sino-Platonic Papers* 177 (August).

De Waal F (2005/6). The evolution of empathy. *Greater good*, 1 September. Retrieved on 30 April 2015 from http://greatergood.berkeley.edu/article/item/the_evolution_of_empathy.

Dibley B (2012). 'The shape of things to come': seven theses on the Anthropocene and attachment. *Australian Humanities Review* 52: 139–53.

Dolphijn R & van der Tuin I (2012). *New materialism: interviews & cartographies*. Ann Arbor, MI: Open Humanities Press.

Doughty A (2008). *An epidemiological survey of the dentition and foot condition of slaughtered horses in Australia*. The University of Queensland, Qld.: School of Animal Studies and the Centre for Animal Welfare and Ethics.

Dougherty C (2013). Videogames about alcoholism, depression and cancer: developers are exploring deeply personal and wrenching stories. *Wall Street Journal*, 15 August. Retrieved on 3 December 2014 from http://online.wsj.com.

Fijn N (2011). *Living with herds: human–animal coexistence in Mongolia*. Cambridge, MA: Cambridge University Press.

Game A (2001). Riding: embodying the centaur. *Body and Society* 7(4): 1–12.

*Gattaca* (2005). Andrew Niccol (Dir). Columbia Pictures, 1997. DVD. Sony Pictures Home Entertainment.

Haraway D (2008). *When species meet*. Minneapolis, MN: University of Minnesota Press.

Hall ET (1983). *The dance of life*. New York: Anchor.

Hempfling KF ([2001] 2011). *Dancing with horses.* London: JA Allen.

Holland M (2013). Former animal rights campaigner Kim Hollingsworth pleads guilty to 11 animal cruelty charges. *Telegraph,* 1 August. Retrieved on 30 April 2015 from http://tiny.cc/aazcyx.

Janzen HH (2011). What place for livestock on a re-greening earth? *Animal Feed Science and Technology* 166–67: 783–96.

Jones G (2013). Clydesdale rescue – life's gentle giants earn a second chance. *Daily Telegraph,* 3 August. Retrieved on 30 April 2015 from http://tiny.cc/7bzcyx.

Joy M (2010). *Why we love dogs, eat pigs, and wear cows: an introduction to carnism.* San Francisco, CA: Conari Press.

Kane B (nd). Cézanne's practice of painting and the ethics of epicureanism Retrieved on 30 April 2015 from http://tiny.cc/hdzcyx.

Kohanov L (2007). *Riding between the worlds: expanding our potential through the way of the horse paperback.* Novato, CA: New World Library.

Lindstrøm TC (2010). The animals of the arena: how and why could their destruction and death be endured and enjoyed? *World Archaeology* 42(2): 310–23.

McGreevy PD, Oddie C, Burton FL & McLean AN (2009). The horse–human dyad: can we align horse training and handling activities with the equid social ethogram? *The Veterinary Journal* 181(1): 12–18.

McManus P & Montoya D (2012). Toward new understandings of human–animal relationships in sport: a study of Australian jumps racing. *Social & Cultural Geography* 13(4): 399–420.

Mullholland S (2011). Externalities. *Learn Liberty,* 29 June. Retrieved on 30 April 2015 from http://www.learnliberty.org/videos/externalities/.

Playthegame.org (2012). Gangs and corruption in Australian horse racing. *Playthegame,* 10 August. Retrieved on 30 April 2015 from http://www.playthegame.org/news/news-articles/2012/gangs-and-corruption-in-australian-horse-racing/.

Price S & Rogers Y (2004). Let's get physical: the learning benefits of interacting in digitally augmented physical spaces. *Computers & Education* 43(1–2): 137–51.

Rollin B 2008. The ethics of agriculture: the end of true husbandry. In MS Dawkins & R Bonney (Eds). *The future of animal farming: renewing the ancient contract* (pp7–19). Oxford, UK: Blackwell Publishing.

Thompson T (2013). Australian slaughterhouses face uncertain future if European horse meat scandal reins in exports. *Courier Mail,* 24 February. Retrieved on 30 April 2015 from http://tiny.cc/cfzcyx.

Tuma KA (2002). Cézanne and Lucretius at the Red Rock. *Representations* 78(1): 56–85.

Gallese V (2003). The roots of empathy: the shared manifold hypothesis and the neural basis of intersubjectivity. *Psychopathology* 36(4): 171–80.

VIZZATA (2012). Consumer perceptions of the horsemeat burger incident. Retrieved on 28 April 2015 from http://www.vizzata.com/docs/vizzata_horsemeat_burgers.pdf.

Wallace J (2005). *DH Lawrence, science and the posthuman.* New York: Palgrave Macmillan.

Wolfe C (2009). *What is posthumanism?* Minneapolis, MN: University Of Minnesota Press.

# 7

# Painfully, from the first-person singular to first-person plural: the role of feminism in the study of the Anthropocene

*Daniel Kirjner*

In 1976, Brazilian poet Thiago de Mello, known for his environmental work, wrote:

> I do not have the sun hidden
> inside my pocket of words.
> I am only a man
> to whom, yet, the first
> and desolated person
> singular – was letting go,
> slowly, painfully,
> from being, to transform itself
> – much more painfully –
> into the first and deep person
> plural.[1]

Kirjner D (2015). Painfully, from the first-person singular to first-person plural: the role of feminism in the study of the Anthropocene. In Human Animal Research Network Editorial Collective (Eds). *Animals in the Anthropocene: critical perspectives on non-human futures*. Sydney: Sydney University Press.

---

1   Thiago de Mello 1989, 3, my translation.

Mello's words are very insightful when thinking about the Anthropocene. They refer to a struggle between his singular conscious existence and a world of social meanings. The pain described by the author is a result of an effort to deconstruct himself as human in order to feel like part of the world: a statement against dichotomous thought that polarises and builds walls between consciousness and existence, nature and society, humans and non-humans.

When it comes to sustainability and environmental issues, these separations are very common and have considerable influence over human behaviour. People are commonly pictured as separate constructs, over and against the idea of nature. But humans and nature are now interacting on a scale unknown before. In the last three centuries human behaviour has affected the so-called natural world in critical ways (Arendt 1998; Crutzen & Stoermer 2000; Filho 2007; Karlsson 2013). Capitalist societies developed so much technologically during this period that their capacity of production and consumption increased as never before: 'Over the past 50 years, humans have changed the world's ecosystems more rapidly and extensively than in any other comparable period in human history' (Steffen et al. 2007, 616). Concurrently, changes in the natural world are affecting human beings.

These cause and effect reactions show that society and nature are intrinsically connected, such that our dichotomous vision of nature versus human must be revisited. Human interactions with non-human animals are a significant part of this equation, and our growing awareness of animal ethics and the environmental impact of animal agriculture has created an upswing of interest in the field of animal studies (Singer 1993; Regan 2001; Patterson 2002). Since about 1990 feminist studies have also turned attention to relations between sexist oppression, speciesism, and environmental degradation (Lorentzen & Eaton 2002). New fields of study (ecofeminism, bioethics, and food cultures) have also focused on interrelations between cultures, non-human animals, and nature (Shiva 1989; Bordo 2003; Diniz 2008).

The concept of the Anthropocene is important for rethinking our dualistic vision and for re-envisioning relations between humanity and nature. This article discusses the role of feminist theory in this work of reevaluation, and the implications of feminism on the study of interconnections between humans, non-human animals, and the environment in the Anthropocene. It is divided in three sections: the first examines the impact of animal exploitation on the environment, the second explores interconnections between ecofeminism and animal studies, and the third discusses the idea of a culture of predation.

## The impact of animal exploitation on the Anthropocene

In the last three centuries the world has experienced a massive environmental upheaval (Chakrabarty 2009). Human influence over nature became more powerful than ever before due to technological advances of the Industrial Revolution (Arendt 1998; Szerszynski 2003).

> The expansion of mankind, both in numbers and per capita exploitation of Earth's resources, has been astounding. To give a few examples: During the past 3 centuries, human population increased tenfold to 6000 million, accompanied e.g. by a growth in cattle population to 1400 million (about one cow per average size family). Urbanisation has even increased tenfold in the past century. In a few generations mankind has nearly exhausted fossil fuels that were generated over several hundred million years. (Crutzen & Stoermer 2000, 17)

Crutzen and Stoermer (2000) used the term 'Anthropocene' to define an era in which humanity's effect on the environment has reached geological proportions. This is not a new idea. Researchers have discussed the implications of large-scale production on the balance of nature since the 18th century.[2] Nonetheless, Crutzen and Stoermer began a fruitful debate focusing on humanity's influence over the Earth.

---

2   See, for example, Steffen et al. 2011; Zalasiewicz et al. 2011. In fact, before the introduction of the concept of Anthropocene, several historical precedents for this far-reaching idea have been revisited. In retrospect, this line of thought, even before the golden age of Western industrialisation and globalisation, can be traced back to remarkably prophetic observers and philosophers of Earth history (Steffen et al. 2011, 844).

Non-humans are painfully affected by cultural changes in the Anthropocene, and also play a prominent role in the Earth's transformations. For example, the world cattle population has more than doubled since the 1960s (World Livestock 2011). Brazil has the largest comercial herd worldwide[3] and the Brazilian Amazonian rainforest is one of the ecosystems most affected by the meat-packing industry.

[T]he majority of deforested areas [in the Amazon] are used for grazing livestock, totaling 77% of the area converted for economic use. To maintain 65 million cattle . . . in the Amazon rainforest, an area of more than 403,000 miles was deforested. (Filho 2007, 161, my translation)

An increase in numbers of cattle has been causing an unprecedented transformation in the Brazilian rainforest. The Amazon is one of the world's main sources of biodiversity, but this proliferation of life has been damaged by deforestation, triggered largely by/for the livestock industry (Filho 2007; Wassenaar et al. 2007; Wilson 2012). In such instances human actions generate abrupt changes in non-human populations. Sometimes the development of one species for animal agriculture triggers the reduction and the extinction of many others (Steffen et al. 2011). A critical change in Earth's biome is one of the main characteristics of the Anthropocene.

A main ingredient of this shift in human manipulation of animals during the last three centuries, is the commodification of life by industrial processes. The use of assembly lines changed the dynamics of the meat industry – work became fragmented, technical, fast and mechanical. At the turn of the 19th century, the meat market was fed by a large, structured industry. Today, one animal is killed every few seconds in most large-scale abattoirs (Patterson 2002).

The use and abuse of animals raised for food far exceeds, in sheer numbers of animals affected, any other kind of mistreatment. Hun-

---

3   More than 209 million cattle heads in 2010, according to the Brazilian Institute of Geography and Statistics (IBGE 2010).

dreds of millions of cattle, pigs, and sheep are raised and slaughtered in the United States alone each year; and for poultry the figure is a staggering three billion. (Singer 2002, 95)

Abattoir technology optimised the number of animals killed per hour, increasing meat production in order to fulfil growing demand in countries like the USA. Some historians, like Charles Patterson (2002), believe that the meat-packing plant was one of the main inspirational models for other contemporary factories (Fitzgerald 2010), following James Barret who suggests that treadmill systems used in packing farms inspired Henry Ford's assembly lines:

> In his study of Chicago's packinghouse workers in the early 1900s, James Barret writes: 'Historians have deprived the packers of their rightful title of mass-production pioneers, for it was not Henry Ford but Gustavus Swift and Philip Armour who developed the assembly line technique that continues to symbolize the rational organization of work.' (Patterson 2002, 72)

The impact of the Industrial Revolution on animal agriculture is an important part of the processes that characterise the Anthropocene. Animal industries are at the root of most major environmental problems, such as deforestation, greenhouse gases, the acceleration of extinction processes (also threatening humans), and the exhaustion of natural resources (Seidl et al. 2001; Filho 2007; Steffen et al. 2007; Chakrabarty 2009; Bowman et al. 2012).

## Ecofeminism and animal studies

François D'Eaubonne, a militant French feminist, forged the connection between feminism and the environment in 1974 with an essay called *Le feminism ou la mort* in which she outlines an ideological basis for ecofeminism (Lorentzen & Eaton 2002). In essence, ecofeminism examines and debates connections between gender oppression and nature's depredation. More specifically, ecofeminism questions the power structure of patriarchy and such society's androcentric dependence on consumption and domination.

By comparing attitudes towards women and nature, ecofeminism explores intersectional violence and questions artificial separations, such as those assumed to exist between nature and culture (Mack-Canty 2004). This tendency to deconstruct traditional Western dichotomies makes ecofeminism an effective tool for analysing the Anthropocene. This geological era, characterised by human imbalance, is defined by the effect that humans have on the environment and the way in which the environment responds. In this context, it is impossible to detach humanity from the rest of the world, or gender issues from environmental problems.

Feminist philosopher Elisabeth Spelman (1982) uses the term 'somatophobia' (aversion to the body) to debate Plato's patriarchal views on the soul and the body. According to her, Plato recognised the equality of men and women in that they possess the same spiritual essence, but goes on to say that the body can corrupt the soul – and women are more affected by bodily influences and the forces of nature than are men. So, the souls of women are more susceptible to corruption, and men are responsible for the control of women's body to elevate their spirits.

> Plato seems to want to make very firm his insistence on the destructiveness of the body to the soul. In doing so, he holds up for our ridicule and scorn those lives devoted to bodily pursuits. Over and over again, women's lives are depicted as being such lives. His misogyny, then, is part of his somatophobia: the body is seen as the source of all the undesirable traits a human being could have, and women's lives are spent manifesting those traits. (Spelman 1982, 118)

Nature is also affected by somatophobia. Plato's association of woman with bodies and nature is symptomatic of patriarchal thinking. The idea that men ought to control women is therefore connected with the idea that men have the right and the duty to overpower nature. The essential existence of humans, according to Plato, is ethereal; the earthly world, including women, must be controlled by this ethereal essence. Therefore sexism is not distinct from environmental degradation, and traditional gender roles must be reexamined if we are to create new relations between humans and the environment (Kheel 1991). In this way ecofeminism is important for understanding the Anthropocene.

Animals, women and the environment are envisioned as closely associated and lesser in this asymmetrical, patriarchal context of power and change, in which female bodies and reproductive systems are appropriated and are genderised. Pioneer Carol J Adams discussed animal exploitation and ecofeminism in her 1990s groundbreaking book, *The sexual politics of meat* (2010), where she compares speciesist violence against animals with sexist subjugation of women. For example, she compares fragmentation of animals into consumable parts (chicken wings, ribs, and so on) with the sexualised fragmentation of women (thighs, breasts, and so on) – all are objects of desire and predatory conquest in patriarchal cultures.

> The animals have become absent referents, whose fate is transmuted into a metaphor for someone else's existence or fate. Metaphorically, the absent referent can be anything whose original meaning is undercut as it is absorbed into a different hierarchy of meaning; in this case the original meaning of animals' fates is absorbed into a human-centered hierarchy. Specifically in regard to rape victims and battered women, the death experiences of animals acts to illustrate the lived experience of women. (Adams 2010, 67)

Adams notes that when a rape victim states that she feels like a 'piece of meat', she is relating graphically and metaphorically to the suffering of animals: she feels broken into pieces, deprived of her complexity, transformed into a consumable fragment of self. Adams refers to this fragmented, exploited individual as an 'absent referent'. Adams' work fosters a productive discussion among ecofeminists (see also Adams & Gruen 2014). Authors such as Josephine Donovan, Greta Gaard, Lisa Kemmerer, and Brian Luke, have since challenged traditional distinctions and preconceptions of species and sex, further developing the field of animal feminist studies (Luke 2007; Gaard 2012; Kemmerer 2011). This debate expands the notion of sexism and gender beyond the limits of humanity. Sexism and speciesism are not different constructs – they are different aspects of the same oppression.

Sexism is speciesist, just as speciesism is sexist. Feminist philosopher Elizabeth Fisher (1979) was one of the first authors to discuss the link between domination of women and domestication of animals:

> The domestication of women followed the initiation of animal keeping, and it was then that men began to control women's reproductive capacity, enforcing chastity and sexual repression. The violation of animals expedited the violation of human beings. (Fisher 1979, 190)

This connection between reproductive control of farmed animals and the domination of women is important for understanding the Anthropocene. The industrial appropriation of female farmed animals' bodies is directly linked to the rise of eugenics and large-scale artificial insemination practices as well as to unsustainable populations of poultry, cattle and swine (Spencer 1995; Patterson 2002; Davis 2005). Females exploited for animal agriculture have a much more controlled existence than do males. They are subjected to a continuous state of pregnancy and lactation that provokes physical exhaustion, intense pain, and that ultimately reduces their life expectancy (Dunayer 1995). It is true that females live longer than their male counterparts, but such a life is not to be envied.

> The cycle of abuse and exploitation for bovine females . . . proceeds for an average of four or five years . . . The bovine's life, outside of these conditions, can be as long as 17 to 25 years . . . This is like sending a young woman to slaughter, after she has given birth eight to ten times . . . between the years of 14 and 25. The 50 years that she could have lived after these births and forced lactations, are not taken into account because her life at this time is no longer commercially useful. (Felipe 2012, 53, my translation)

Although the rape of cows in animal agriculture is different from the rape of women in some ways, farm animals are not objects, and their sexual exploitation therefore cannot go unquestioned. Cows are sentient and socially complex, and awareness of sexism is important for understanding the ethical implications of reproductive control in animal agriculture in patriarchal societies. Domestication and sexual repression are historically connected (Fisher 1979).

Maria Comninou (1995), Carol J Adams (2004) and Brian Luke (2007) argue that masculine values are directly related to the exaltation of violent, predatory behaviour. Aggression and dominance are proud signs of virility, as opposed to compassion and empathy, which are diminished as feminine. These patriarchal attitudes and values affect both women and other animals.

> Within animal exploitation itself the gender differentiation is even more apparent. Throughout the twentieth century well over 90 percent of North American hunters have been male. 'Man the Hunter' is a long-standing and highly entrenched image, and this affects our perception of other institutions of animal exploitation, such as meat production . . . Such associations, along with the continuing stereotype of the macho cowboy driving livestock to the slaughter, support a deep cultural linkage between meat and manhood . . . (Luke 2007, 12)

It is important to recognise the Anthropocene as stemming from a sexist, speciesist, predatory, racist culture (Harper 2010). Only by exposing otherwise hidden assumptions of patriarchal societies, such as dualism and linked oppressions, is it possible to understand the Anthropocene. In 1975, ecofeminism was just emerging, yet a couple of ideas were already clear: (1) feminists and environmentalists ought to cooperate; (2) they ought to unite in opposition to ideologies and values of domination.

## A culture of predation

Ecofeminist Karen Warren, in her book *Ecofeminist philosophy*, credits Rosemary Radford Ruether with changing her academic life, citing this quote:

> Women must see that there can be no liberation for them and no solution to the ecological crisis within a society whose fundamental

model of relationships continues to be one of domination. They must unite the demands of the women's movement with those of the ecological movement to envision a radical reshaping of the basic socioeconomic relations and the underlying values of this society. (Ruether 1975, 204)

Although domination is an expansive idea, Rosemary Radford Ruether referrs to a particular type of domination that affects both women and nature. Many ecofeminists reference an oppressive, predatory power that is associated with gender roles, calling attention to connections between rationality, denigration of women and nature, and capitalism (Warren 2000).

Modern science was a consciously gendered, patriarchal activity. As nature came to be seen more like a woman to be raped, gender too was recreated. Science as a male venture, based on the subjugation of female nature and female sex provided support for the polarization of gender. Patriarchy as the new scientific and technological power was a political need of emerging industrial capitalism. (Shiva 1989, 17)

Ecofeminist Vandana Shiva associates the capitalist exploitation of nature with the culture of rape. The Western commitment to rationality is based on somatophobic thinking where reason is imbued with spiritual significance while the physical world is conquered and is commodified. In this context not only women and animals but the environment as a whole becomes an 'absent referent'.

The absent referent, as conceptualised by Carol Adams, is a very important tool for understanding cultures of predation and the Anthropocene. Fragmentation makes violence palatable because the victim disappears. Emotional detachment is necessary for domination. 'Beef' is not a cow just as a fragmented body part is not a woman, and a '2x4' is not a tree. In capitalist societies, just as farmed animals are fragmented to become meat, and thereby disappear as living, complex beings; and just as women are fragmented to become sexual objects, and thereby disappear as complex individuals; so is nature fragmented and denied in the industrial processes. Capitalist products fragment and commodify the environment in a manner that often prevents consumers from realising the impact of production and of their consumption. For example, consider corn:

> There are some 45000 items in the average American supermarket and more than a quarter of them now contain corn. This goes for the non-food items as well everything from toothpaste and cosmetics to disposable diapers, trash bags, and even batteries. (Pollan 2006, 11)

In most of these products, corn is invisible. Few consumers know how often they buy corn, or the impact of the consumption of this product on the environment. Fragmented and packaged for consumption, corn is further disguised by incomprehensible scientific names and minute print strategically placed on the edges or bottoms of a package. Corn, like other commodified aspects of the natural world, is a processed, packaged fragment of nature, no longer easily identified for what it once was. Such products stem from violence, and domination, over nature.

Predation, a hunter feeding on prey, entails power and violence as the hunter overcomes the will of others. In the act of killing, conservation and balance are not valued; all that matters is the satisfaction of the whims and pleasures of the stronger without regard to the most basic interests of the weaker. Patriarchal cultures tend to value predation. Conquest is more glamorous than balance. Masculinity is bolstered through the act of violent domination. In predatory cultures, women, animals and nature are all prey, prizes to be won, conquests, an affirmation of identity through power-over. Both women and animals become

> psychological instruments to the establishment of the masculine self ... animals are still used as instruments of self-definition; they are killed not in the name of an individual masculine ego, but instead in the name of a higher, abstract self. (Kheel 1991, 69)

The Anthropocene has emerged in an atmosphere in which males are encouraged to be predators, and in which women, animals and nature are prey. In this sense, environmental transformations that characterise the Anthropocene cannot be considered apart from norms of masculinity. Ecofeminists Carolyn Merchant, Val Plumwood and Maria

Mies elucidate relations between capitalism, technology and masculinity – what I am calling a culture of predation (Merchant 1989; 2014; Mies 1998; Warren 2000; Plumwood 2002).

The Anthropocene, shaped by a culture of predation, is characterised by fragmentation, absent referents and violent domination. It is not a coincidence that predation is largely male-centred. Hunters in the USA are almost all male; animal agriculture and slaughter are largely dominated by men (Kheel 1996; Luke 2007; Cudworth 2008; Parry 2010; Graf & Coutinho 2012). The reproductive abilities of farmed animals are exploited to multiply these species to unsustainable proportions, while other species are endangered or are forced into extinction. Women are dismembered into sexual fragments and are domesticated, reaffirming masculine power and control (Fisher 1979; Adams 2004). The environment is fragmented and is exploited as a commodity under the power of capitalism. These are all indicators of the interconnected nature of sexism, speciesism, environmental degradation, and masculinity in patriarchal cultures. The Anthropocene appears to be a logical outcome of a culture of predation.

## Conclusion

The Anthropocene exemplifies the destructive effects of patriarchal culture on the environment. Ecofeminist perspectives provide compelling tools for analysing the centrality of gender roles to these significant environmental changes. Environmental hazards threaten the status quo as never before. Patriarchal, dichotomous thought is too rigid and too narrow to cope with the current velocity and magnitude of change.

The oppression of animals in the Anthropocene is certainly not justified by biological differences between humans and non-humans. Animal exploitation stems from a reframing of life caused by capitalist commodification, by a predatory culture that glorifies male violence, and by the fragmentation and stigmatisation of emotion as a lesser form of expression than cold rationality and violent domination.

Recalling Thiago de Mello's verses we are reminded that the sun cannot be hidden in a pocket; its light is everywhere. We cannot live by predatory, individualistic desires, without a rebellion from nature. The first-person singular is a transitional and painful state heading to an always-incomplete first-person plural. The Anthropocene characterises itself not as an era of dichotomies and individualities but as one of inescapable intersections.

## Acknowledgment

I would like to thank Professor Lisa Kemmerer for having carefully and attentively revised this work. Without her hard work, this chapter would not have been possible.

## Works cited

Adams CJ (2004). *The pornography of meat*. New York: Bloomsbury Academic.
Adams CJ (2010). *The sexual politics of meat*, 3rd edn. New York: Continuum International Public.
Adams CJ & Gruen L (2014). *Ecofeminism: feminist intersections with other animals and the earth*. London: Bloomsbury.
Arendt H (1998). *The human condition*, 2nd edn. Chicago, IL: University of Chicago Press.
Bordo S (2003). *Unbearable weight*. Los Angeles, CA: University of California Press.
Bowman MS, Soares-Filho BS, Merry FD, Nepstad DC, Rodrigues H & Almeida OT (2012). Persistence of cattle ranching in the Brazilian Amazon: a spatial analysis of the rationale for beef production. *Land Use Policy* 29(3): 558–68.
Chakrabarty D (2009). The climate of history: four theses. *Critical Inquiry* 35(2): 197–222.
Comninou M (1995). Speech, pornography and hunting. In C Adams & J Donovan (Eds). *Animals and woman: feminist theoretical exploration* (pp126–47). Durham, NC: Duke University Press.
Crutzen P & Stoermer EF (2000). The 'Anthropocene'. *Global Change Newsletter* 41: 17–18. Retrieved on 29 April 2015 from http://www.igbp.net/download/18.316f18321323470177580001401/NL41.pdf.
Cudworth E (2008). 'Most farmers prefer blondes': the dynamics of anthroparchy in animals becoming meat. *Journal for Critical Animal Studies* VI(1): 32–45.
Davis K (2005). *The Holocaust and the henmaid's tale: a case for comparing atrocities*. New York: Lantern Books.

De Mello T (1989). *Poesia comprometida com a minha e a tua vida*, 6th edn. Rio de Janeiro: Bertrand Brasil.

Diniz D (2008). Bioética e gênero. *Revista Bioética* 16(2): 207–16. Retrieved on 28 April 2015 from http://revistabioetica.cfm.org.br/index.php/revista_bioetica/article/view/68.

Dunayer J (1995). Sexist words, speciesist roots. In C Adams & J Donovan (Eds). *Animals and woman: feminist theoretical exploration* (pp11–30). Durham, NC: Duke University Press.

Felipe ST (2012). *Galactolatria: o mau deleite*. São José: Ecoânima.

Filho JM (2007). *O livro de ouro da Amazônia*, 5th edn. Rio de Janeiro: Ediouro.

Fisher E (1979). *Woman's creation: sexual evolution and the shaping of society*. Garden City, KS: Anchor Press.

Fitzgerald AJ (2010). A social history of the slaughterhouse: from inception to contemporary implications. *Human Ecology Review* 17(1): 58–69.

Gaard GC (2012). Feminist animal studies in the US: bodies matter. *Revista Tellematica di Studi Sula Memoria Feminile* (20): 14–21.

Graf LP & Coutinho MC (2012). Entre aves, carnes e embalagens: divisao sexual e sentidos do trabalho em abatedouro avicola. *Revista Estudos Feministas* 20(3): 761–83.

Greenpeace (2009) *O rastro da pecuária na Amazônia*. São Paulo: Greenpeace.

Harper AB (Ed) (2010). *Sistah vegan*. New York: Lantern Books.

IBGE (2010) *Produção da Pecuária Municipal 2010*. Rio de Janeiro: Instituto Brasileiro de Geografia e Estatística.

Karlsson R (2013). Ambivalence, irony, and democracy in the Anthropocene. *Futures* 46: 1–9.

Kemmerer L (2011). *Sister species: women, animals and social justice*. Champaign, IL: University of Illinois Press.

Kheel M (1991). Ecofeminism and deep ecology: reflections on identity and difference. *The Trumpeter: Journal of Ecosophy* 8(2): 62–72. Retrieved on 28 April 2015 from http://trumpeter.athabascau.ca/index.php/trumpet/article/view/784.

Kheel M (1996). The killing game: an ecofeminist critique of hunting. *Journal of the Philosophy of Sport* 23(1): 30–44.

Lorentzen LA & Eaton H (2002). Ecofeminism: an overview. Retrieved on 28 April 2015 from http://skat.ihmc.us/rid=1174588237625_665601541_9501/ecofeminism.pdf.

Luke B (2007). *Brutal: manhood and the exploitation of animals*. Urbana, IL: University of Illinois Press.

Mack-Canty C (2004). Third-wave feminism and the need to reweave the nature/culture duality. *NWSA Journal: A Publication of the National Women's Studies Association* 16(3): 154–79.

Merchant C (1989). *The death of nature: women, ecology, and the scientific revolution*. New York: Harper & Row.

Merchant C (2014). *Earthcare: women and the environment*. London: Routledge.

Mies M (1998). *Patriarchy and accumulation on a world scale: women in the international division of labour*. London: Atlantic Highlands.

Parry J (2010). Gender and slaughter in popular gastronomy. *Feminism & Psychology* 20(3): 381–96.

Patterson C (2002). *Eternal Treblinka: our treatment of animals and the Holocaust*. New York: Lantern Books.

Plumwood V (2002). *Environmental culture: the ecological crisis of reason*. New York: Routledge.

Pollan M (2006). *The omnivore's dilemma: a natural history of four meals*. New York: Penguin Press.

Regan T (2001). *Defending animal rights*. Champaign, IL: University of Illinois Press.

Ruether RR (1975). *New woman, new earth: sexist ideologies and human liberation*. Melbourne, Vic.: Dove Communications.

Seidl AF, De Silva JDSV & Moraes AS (2001). Cattle ranching and deforestation in the Brazilian Pantanal. *Ecological Economics* 36(3): 413–25.

Shiva V (1989). *Staying alive: women, ecology and survival in India*. London: Zed Books.

Singer P (1993). *Practical ethics*, 2nd edn. Cambridge, UK: Cambridge University Press.

Singer P (2002). *Animal liberation*, 3rd edn. New York: HarperCollins Publishers Inc.

Spelman E (1982). Woman as a body. *Feminist Studies* 8(1): 109–31.

Spencer C (1995). *The heretic's feast: a history of vegetarianism*. Hanover, NH: University Press of New England.

Steffen W, Crutzen PJ & McNeill JR (2007). The Anthropocene: are humans now overwhelming the great forces of nature? *A Journal of the Human Environment* 36(8): 614–21.

Steffen W, Grinevald J, Crutzen P & McNeill J (2011). The Anthropocene: conceptual and historical perspectives. *Philosophical Transactions of the Royal Society A: Mathematical, Physical and Engineering Sciences* 369(1938): 842–67.

Szerszynski B (2003). Technology, performance and life itself: Hannah Arendt and the fate of nature. *Editorial Board of the Sociological Review* 52(s2): 203–18.

Warren K (2000). *Ecofeminist philosophy: a Western perspective on what it is and why it matters.* Lanham, MD: Rowman & Littlefield.

Wassenaar T, Gerber P, Verburg PH, Rosales M, Ibrahim M & Steinfeld H (2007). Projecting land use changes in the neotropics: the geography of pasture expansion into forest. *Global Environmental Change* 17(1): 86–104.

Wilson EO (2012). *Diversidade da vida.* São Paulo: Companhia das Letras.

World Livestock (2011). *Livestock in food security.* Rome: Food and Agriculture Organization.

Zalasiewicz J, Williams M, Haywood A & Ellis M (2011). The Anthropocene: a new epoch of geological time? *Philosophical Transactions of the Royal Society A: Mathematical, Physical and Engineering Sciences* 369(1938): 835–41.

# 8

# We have never been meat (but we could be)

*Simone J Dennis and Alison M Witchard*

Throughout history, humans have required access to flesh to gain food, health and knowledge. Our access to flesh is made on the basis that it is significantly different from human flesh: we do not, by and large, consume our own flesh, but instead that of the animal other. This hierarchically arrayed difference is the basis upon which humans can enact culturally acceptable violence upon the body of the animal, to produce the fleshy commodification of its parts. As Nick Fiddes (1991, 277) suggests of flesh accessed for food, meat's essential value 'derives directly from its capacity to represent to us most tangibly our power over the rest of the natural world' (see also Chakrabarty 2009, 209).

This flesh has been recognisably different from 'us' because of its containment within a whole animal body – a pig, a cow, a lamb. But a new biotechnology, in-vitro meat, does away with the need for a whole animal body that serves as the genitor of flesh: it uses cells. This technology, we argue, also does away with human–animal difference. Presently, the cells that in-vitro meat technology uses are taken from an animal donor, but there is no technical barrier that would

Dennis SJ & Witchard AM (2015). We have never been meat (but we could be). In Human Animal Research Network Editorial Collective (Eds). *Animals in the Anthropocene: critical perspectives on non-human futures*. Sydney: Sydney University Press.

prevent human progenitors from being the cellular input into in-vitro meat technology. This possibility cuts at the key tenets of hierarchical species difference that currently enable the human consumption of animal flesh. Like other changes that have undermined the surety of the Anthropocene, including the ramifications of human-made climate change, the possibility of in-vitro meat that uses human cellular inputs means that human domination dissolves into a future that we cannot visualise, since that future is no longer premised on the domination of nature and animals (Chakrabarty 2009, 211). The possibility of a world cognisant of the second law of thermodynamics, and the possibility of the cessation of the instrumentalisation of animals does not mean that either of these things will come to pass; the future of the biotechnology is very unclear. It may usher in a utopia, as groups such as People for the Ethical Treatment of Animals (PETA) hope, in which animal flesh is no longer used by humans for gain, and where they stand equal. Just as possibly, we could be left with a situation in which all beings emerge as equally open to commoditisation – each a potential source of flesh, none able to 'witness the inhuman' rendered unattainable (Deranty 2008). We do not know; but what we do know is its potential for reframing how we think about animal–human relations.

## Meat: a key marker of human and animal difference

In the West, meat appropriate for human consumption is rendered so under exacting circumstances. An animal body, the bearer of marked and observable species difference (that is by its bearing of hoofs and horns, feathers and fur) is submitted to a range of industrialised (and hidden) processes: stunning, bleeding, dismantling, and packaging that render a whole animal body into consumable parts. This arrangement, symptomatic of the Anthropocene, is entrenched as much in the gathering of meat as it is in the teachings of the Bible, although they are differently arrayed in their varying emphases on killing and sacrifice respectively. The evident difference of these bodies from our own, and the hierarchical processes in which they are thus entailed, are crucial to creating the fleshy substance called meat. As Cora Diamond (1978, 470) suggests, '[w]e learn what a human being is – among other ways – sitting at a table where *we* eat *them* [animals]. We are around the table and they are on it' (our italics).[1]

How animals get onto the table has changed markedly. In pre-industrial Europe, animal slaughter was a highly visible presence within everyday life (Fiddes 1991; Vialles 1994; Fitzgerald 2010). Often conducted on the streets of cities and towns, the slaying and flaying of an animal was considered an ancient practice or artisan craft (Brantz 2001). Animals were slaughtered in public using methods passed down through many generations. Such processes detailed by Mercier (1782):

> a young bull is thrown down and his head is tied to the round with a rope; a strong blow breaks his skull, a large knife gives it a deep wound in the throat; streaming blood spills out in big bursts along with the life ... Bloody arms plunge into its steaming innards, a blowpipe inflates the expired animal and gives it a hideous shape, its legs are chopped off with a cleaver and cut up into pieces and at once the animal is stamped and marketed. *(Tableau de Paris*, Vol. 5, pp101–3, quoted in Brantz 2001, np)

The visceral immediacy of this form of slaughter is expressed above, as one or two men attempt to butcher, unassisted by complex machinery, the living, struggling animal body as a whole, and dismantle it entirely. The subsequent need for swiftness in distributing and in consuming meat, given the lack of refrigeration or modern preserving techniques, resulted in a specific set of qualities that suggested 'good' or desirable meat, and the emergence of particular forms of human–animal relationships. The 'warmth' and bloodiness of animal products, such as

---

1   Within this analogy, there also exists the possible presence of another animal; the household pet, under the table waiting for the scraps: it is important to recognise that this animal is uniquely situated and is assigned a 'subjectivity' quite different from the livestock that provide our meat. This pet is both different from the person, it belongs to another 'species', so could technically qualify as meat. However, this pet, assigned a name, provided with care often on par to kin, lingers too close to the person; this dog, cat and so on cannot taxonomically qualify. Animal referred to in this chapter is a specific animal or 'animal' subjectivity, an animal deemed suitable for consumption within Western society.

milk and meat, Richie Nimmo (2011, 60) notes, was considered indicative of fresh and authentic animality. The 'extraction of animal traces' from meat products as occurs currently, he continues (Nimmo 2011, 61), was not viewed so positively in the past as it is today. Conversely, the perceivable proximity and presence of the animal life was considered desirable. These traces of life represented the 'material flows of energy' from the warm blood and fleshy cuts of animal, this vitality to be transferred to the individual through digestion (Nimmo 2011, 61). For example, Nimmo (2011, 62–63) notes that, prior before 1985, milk was considered:

> *part of* the cow, a product of its body, and as such, inseparable from its mode of species life and its fleshy bovine being. The cow, or rather, cows collectively, were very much materially and ontologically present in the milk, and the consumption of milk was a human–bovine encounter in a quite immediate sense. Moreover, the milk itself constantly testified to its 'cowness' by its very inconsistency and perishability, and by the everyday visibility of the cows in the urban cowsheds which were necessarily physically proximate to the places where the milk was consumed ... [I]t was an [human–animal] encounter fundamentally rooted in proximity and presence rather than distanciation and mediation.

With the advent of the modern scientific era throughout Europe, the popularity of meat as an essential source of material to 'replenish muscular strength' prospered (Liebig 1846 in Fiddes 1994, 275), and with it a social reform movement to increase the amount of meat available to all classes of society, particularly those who, as advocate William Cobbett reported, 'raise all of this food' (Knapp 1997, 545).[2] This 'de-

---

2 While meat remained the purview of those wielding social power and control during early history, with the advent of new farming techniques and technologies, such as the more efficient types of plough, the amount of feed available to raise livestock increased, resulting in what historian Fernand Braudel described as the 'l'Europe carnivore'. By the late Middle Ages, Europe emerged as the most meat-eating culture in the world, with larger quantities of meat consumed mainly by the upper classes, although its production remained the purview of the lower (Braudel in Knapp 1997, 540). Vegetarian diets, at least in Europe, were still

mocratisation of meat' resulted in a change in the processes applied to animal bodies that yield meat, ushering in new technologies and forms of human management and control of animals on an industrial scale, firmly placing humans and animals, culture and nature, killers and victims, on either side of a firm hierarchical divide (Chakrabarty 2009).

## Janus flesh

One major change attending the democratisation of meat and industrialisation in Europe was the urbanisation of animal slaughter. In her ethnography of French abattoirs, Noelie Vialles (1994) charts the movement of animal slaughter out of the cities and into new designated and controlled spaces, such containment enabling the monitoring and mechanisation of animal slaughter and disassembly. The need for mass killing of animals on a scale hitherto unseen produced different conditions for animal slaughter, radically changing the spatiality of slaughter, as large abattoirs opened up outside of cities such as Paris and Chicago's Union Stock Yard. With this shift, the grandeur and horror of the Middle Ages was past; 'the charnel house of the Holy Innocents with its mud-caked skulls in open-air loggias amid stalls of fruit and meat and dish; hanged criminals in the small of the tide; religious banners being paraded among tubs of entrails'. To comply with the new standards of hygiene and the increased demand for meat products, private slaughter was rendered prohibited, as the killing of animals was required to be industrial; 'large-scale and anonymous' (Vialles 1994, 22).

As a result of these shifts, meat in the industrialised West is now something other than what it once was. Now, it must be 'certifiably safe, hygienic, traceable, ethically acceptable and of a standardised cut' (Reinert 2007, np) and essentially, as Vialles (1994) notes, '*de*-animal-

---

viewed as the 'poor peasant relation', with meat consumption remaining a matter of prestige before 1800. Many peasants would go a year without consuming meat, as British aristocrat Sir William Petty noted in 1690s '[a]s for flesh, they seldom eat it' (in Knapp 1997, 543). During those times, however, aristocrats continued to advocate a 'meat-heavy diet'. Meat-eating, however, was not everywhere symptomatic of the upper echelons of society. For example, within the Hindi culture, in contrast, the elite Brahmin cast refrain from eating meat (Harris 1966).

ized'. Nimmo (2011) describes a similar shift occurring in the milk industry at the turn of the 20th century. With the advent of microbiology and high public hygiene concerns, 'the tangible trace of the living animal [the 'cow-ness' of the milk] becomes something deeply problematic'. Like meat, milk was divorced from its animal corporeality; however, such a transition into a manufactured commodity was by no means simple or straightforward. 'Whilst it could be crossed out', animal materiality of these products 'could by no means be erased, and while it could be held in abeyance it could not be made to disappear' (Nimmo, 2011, 66). A similarly 'Janus-esque' situation attends the making of meat: society demands both the creation and the destruction of animal being under human hand (Bauman 1989).

To be meat, flesh must come from an animal being slaughtered under human control however it must also reduce any other undesirable qualities that point to the act: blood, fur, hair, hoofs. Consequently, we know where our meat comes from, we would, however, prefer not to be reminded of it. Such a paradoxical situation is encapsulated by Vialles (1994, 51), who notes that meat can be defined as,

> the homogenous substance, all of a piece, that results from the elimination of the animal parts, whether external or internal, that gave the former *whole* its visible autonomy (hide, head, feet) or were the secret source of its animation (heart, lungs, liver, guts). An animal (in the ordinary sense of an animated creature) is from the point of view of consumption simply a machine for manufacturing flesh; once the product has reached maturity and been harvested ... dressing releases the meat from its biological apparatus, the sole purpose of which was to secrete it. The whole dressing process is thus in fact an *undressing*, not only in that it removes the animal's external envelope but above all in that it strips the flesh of its animality, detaching the organic substance from its biological foundations: meat is an organic substance obtained by dispersal of the biological [our italics].

This demand of an animal progenitor for meat and the equal need for de-animalisation, for it to count as a consumable product, Fiddes (1994, 277) suggests, demonstrates the 'severity of the spiritual divorce we ordain between ourselves and the world that supports us'. Such bifurcation is characteristic of modern Western society. Thus '[a]nimals in name and body', Adams (1990, 10) argues, 'are made absent as animals for meat to exist. Animals' lives thus essentially precede and enable the existence of meat. If animals are alive, they cannot be meat.' Animals are then, as Reinert (2007, np) notes, the 'represented presence in the meat', a 'public secret of meat consumption'.

The rise of industrial slaughter had very real effects not only on the techniques, processes and materiality of meat but also on the relationships, sentiments, politics and pragmatics surrounding these seemingly mundane methodologies of modernity. The abattoir, Reinert (2007, np) asserts, with its play between 'concealment and knowledge, suspicion and fact', has emerged as a 'powerfully and morally charged trope of the modern Western imagination, as well as the most immediately recognisable element of a modern animal industrial complex'. The modern animal-industrial complex, Twine (2012, 22) asserts, operates largely to quell the 'potential disruptive potency' of violent animal slaughter to call into question the 'naturalisation' of the human–animal hierarchy.

## We have never been meat (but we could be)

But a biotechnology that arises outside of the animal–human hierarchy might cause just as much disruption – even more than the discomfort that comes with enacting lethal violence against animal, and covering it over. In-vitro meat, a new biotechnology able to produce 'meat' tissue grown from, potentially, cells extracted from any animal – chicken, pig, cow, komodo dragon, panda, humans, has caused much controversy since its announcement in the early 2000s.

A single cell is extracted from any organic donor, without necessitating its death, and is then grown and is exercised in nutrient-rich fluid within the laboratory to produce flesh. It will eventually emerge as a lump of unidentifiable flesh, yet without bodily bounded life, without blood, without the markers of the species from which it originated, its origins at input fundamentally unimportant to the technology's ability to produce flesh.

The biotechnology of in-vitro meat represents the cutting edge of current scientific developments seeking to address many of the most pressing global, social and ecological challenges of the Anthropocene, issues such as food insecurity, global warming, pandemics, including the recent H1N1 swine flu, ecological degradation and most prominently, animal abuse and misuse. The marketing of the developing in-vitro meat technology speaks directly to a number of these problems. President of PETA, Ingrid Newkirk is an enthusiastic proponent of in-vitro meat's capabilities, declaring 'what a joy [it will be] to be able to give them animal flesh that comes without the horror of the slaughterhouse, the transport truck, and the mutilations, pain, and suffering of factory farming'. The resulting 'meat' substance or tissue culture is theoretically capable of revolutionising current and increasingly problematic modes of meat production and distribution according to one of its key developers, vascular biologist at Maastricht University, Mark Post (in Krijnen 2012; see also Post 2012),

> Current livestock meat production is just not sustainable, not from an ecological point of view, and neither from a volume point of view . . . It's simple maths. We have to come up with alternatives . . . With cultured meat we can be more conservative – people can still eat meat, but without causing so much damage.

In 2008, PETA offered a $1 million prize to the first group of scientists able to produce marketable in-vitro meat; the deadline for this prize was extended to 4 March 2014. When it too expired without award, PETA remarked on its website that the original prize had been for the production of a commercial quantity of in-vitro chicken meat, since 'chickens are the most abused individuals used for food by virtue of their sheer number, with a million slaughtered in the USA alone each hour'. PETA also used the expiration of the deadline to declare that the announcement of a prize in any case, even if it had not been claimed, was yet instrumental: 'Since we announced the prize, laboratory work on in vitro meat has come a long way, and a commercially viable beef hamburger or pork sausage are bound to happen in the not-too-distant future' (PETA 2015).

For PETA, in-vitro meat presents the best of both worlds. Animals are released from relationships of routine slaughter and commercialisation and the 'many people who continue to refuse to kick their meat addictions . . . [can] gain access to flesh that doesn't cause suffering and death' (PETA 2011). In this future world, humans and animals emerge as laterally related, even equal, each afforded the same right to life as the other. In PETA's view, in-vitro meat thus offers a means of undoing the present disastrous hierarchical arrangements between humans and animals, perhaps helping to realise Szerszynski's prediction for the Anthropocene as the 'Epoch of the apotheosis, or of the erasure, of the human master and end of nature' (2010, 16 in Dibley 2012, 141).

With animal rights groups like PETA, some ethicists, philosophers, and members of the public support the development of in-vitro meat. However, the overwhelming response to in-vitro meat is that of revulsion and of disapproval. In-vitro meat is commonly labelled as 'unnatural', 'franken-meat', 'zombie-meat', and as having an insurmountable yuck factor sufficient to prevent people from eating it (Hopkins & Dacey 2008; Stephens 2010). Such sentiments are encapsulated in the following responses to PETA's announcement:

> This is disgusting and reprehensible. If you do not want to kill animals for meat then do not eat meat. If you are worried about the humane treatment of animals then fight that fight. But creating test tube meat in order to reach both of these goals is an idiotic pursuit for such an organization. (Hyena 2009, np)

Such responses might seem unfathomable; why would people disapprove of a technology that might emancipate animals from the lethal violence enacted against them, and that might emancipate us from the environmental consequences of producing meat? A number of responses have already been proposed. Public unease towards in-vitro meat has been explored along such lines as risk and danger (Hopkins & Dacey 2008), naturalness/unnaturalness (Stephens 2010; Hopkins & Dacey 2008), and a fear of unknown technological manipulation (as Tim's comment, about 'test-tube meat' suggests – see Stephens 2010). But none of these directly engages the fact that:

> In vitro meat will be fashioned from any creature, not just domestics that were affordable to farm. Yes, ANY ANIMAL, even rare beasts like snow leopard, or komodo dragon. (Hyena 2009, np)

Picking up on the tone of public opinion circulating about in-vitro meat, food ethicist, Chris MacDonald (2009) commented that people are 'worried *not* just about "plain" *in vitro* meat, but about the creations the *in vitro* meat process might enable'. A similar argument propounding the need for strict parameters surrounding in-vitro meat technology is offered by Gillian Madill, Genetics Technologies spokesperson for Friends of the Earth, a non-for-profit organisation dedicated to socially equitable and environmentally sustainable futures:

> If we can successfully develop these products, what is the defining line between lab-grown meat and natural animals? That is an especially important question since a high level of differentiation and tissue complexity is required to replicate muscle tissue that we use as meat. We need to draw clear lines in order to prevent the commodification of all life. (Ford 2009)

There is another possibility for why alarm is being sounded. We submit that such discomforts arise from the possibility of the undermining the certain and 'normal' conditions of the Anthropocene – the exploitation of animal bodies for human gain. So entrenched has this modus operandi become – to science, to health, to sustaining our lives calorically – that undermining it presents us with an unthinkable world in which we might be reduced to being eaten – *we might be meat.*

To contemplate the potential of in-vitro meat to violate our accepted norms and cherished categories, such as those of difference between humans and animals, between nature and culture, requires consideration of the technology's capabilities. Able to utilise any cell: animal, human, even plant, it holds a unique ability to erode our social 'certainties', dispelling with the hierarchical relationships between human and animal we currently experience, particularly in terms of meat-eating. Not a certainly and definitively human or animal generator in-vitro meat technology jettisons a number of our most cherished and axiomatic classifications that bolster social–material order; humans as exceptional, animals as instruments or sources of meat, as we all emerge as potential progenitors of this new form of flesh. Haraway, in her *Cyborg manifesto* ([1985] 2004, 24), suggests that biotechnologies blur the difference between organism and machine; subsequently 'mind, body and tool are on very intimate terms'. We suggest that in-vitro meat technology goes a step further. The processes currently considered necessary for producing 'real' meat: the rearing of an animal, its slaughter, the spilling of its blood, skinning and disembowelling of its natural body, and the packaging and marketing of its parts as 'desirables'is replaced by the careful 'culturing' of cells within a petri dish. Thus, instead of merging the categories of human and animal, as is the case with many biotechnologies, in-vitro meat does away with them altogether by not insisting that a particular kind of cell is needed to make meat, thus eliminating a key distinction between animals and humans and a maintainer of that binary – *now everyone could be meat.*


## Conclusion

Biotechnologies culminate in what Nikolas Rose (2007) describes as a 'new ontology of life' taking place at the molecular level, and thus, we propose, largely devoid of the markers of the categories of human and animal. This move, Rose (2007) argues, entails a new attention towards the 'molecular' rather than the 'molar', in doing so it provides an alternative 'scalar narrative' crucial to our understandings of life in the Anthropocene, forcing us to reflect on who we are and what we are made of as our 'species' thinking and history comes under revision. Such shifts demonstrate what Paul Rabinow (1999, 16) describes as the problematisation of 'life' as 'new understandings and technologies that are involved in giving it a form are producing results that escape the philosophical self-understandings provided by both the classical world and the Christian tradition' – these traditions including the Cartesian divide between humans and animals.

With the advent of in-vitro meat life is no longer conceptualised as a macro-anatomical system, a bounded and enclosed whole embedded in taxonomic systems of binary oppositions, but rather emerges as manipulable by the technologies of in-vitro meat. While we may continue to conceptualise our bodies and being-ness on a 'molar' level, Rose (2007, 5) suggests that this scale is increasingly being replaced by life as understood and enacted on the molecular level. Biotechnology enables the 'decomposing, anatomising, manipulating, amplifying, reproducing' of vitality in genes, cells, tissues and so forth. These elements, once considered the purview of each individual body, are now accorded a new mobility as seen in in-vitro meat.

Able to be isolated and extracted from the bodily whole, the cell – be it human or animal – emerges as able to exist in vitro, free from visible indicators of its progenitor; it suggests life's vitality, its interchangeability, and its capacity to trouble the divisions that insert themselves between different kinds of human and animal bodies. In the absence of whole, distinguishable bodies, there persists the cell. Drawn from any body, yet itself bodiless, it effortlessly crosses the lines that have separated life from death, Other from us, dead meat from living killers, animals from humans. In-vitro meat is in its development stage; it is an object about which we know little in terms of its circulation and meaning in the world. Yet one thing we can say with certainty is that the rupture to the animal–human divide presented by such technology will indubitably produce (perhaps fruitful) insights into what it means to be a 'being'.

## Works cited

Adams C (1990). *The sexual politics of meat: a feminist-vegetarian critical theory.* New York: Continuum.

Bauman Z (1989). *Modernity and the Holocaust.* New York: Cornell University Press.

Brantz D (2001). Recollecting the slaughterhouse. *Cabinet* 4. Retrieved on 30 April 2015 from http://cabinetmagazine.org/issues/4/slaughterhouse.php.

Chakrabarty D (2009). The climate of history: four theses. *Critical Inquiry* 35: 197–222.

Deranty J (2008). Witnessing the inhuman: Agamben or Merleau-Ponty. *South Atlantic Quarterly* 107(1): 165–86.

Diamond C (1978). Eating meat and eating people. *Philosophy* 53: 465–79.

Dibley B (2012). 'The shape of things to come': seven theses on the Anthropocene and attachment. *Australian Humanities Review* 52: 139–53.

Fiddes N (1991). *Meat, a natural symbol.* New York: Routledge.

Fiddes N (1994). Social aspects of meat eating. *Proceedings of the Nutrition Society* 54(2): 271–79.

Fitzgerald A (2010). A social history of the slaughterhouse: from inception to contemporary implications. *Research in Human Ecology* 17(1): 58–69.

Ford M (2009). In-vitro meat: would lab burgers be better for us and the planet? *CNN News*, 8 August. Retrieved on 30 April 2015 from http://www.cnn.com/2009/TECH/science/08/07/eco.invitro.meat/index.html.

Haraway D (2004). *The Haraway reader*. London: Routledge.

Harris M (1966). The cultural ecology of India's sacred cattle. *Current Anthropology* 7(1): 51–66.

Hopkins PD & Dacey A (2008). Vegetarian meat: could technology save animals and satisfy meat eaters? *Journal of Agricultural and Environmental Ethics* 21(6): 579–96.

Hyena H (2009). Eight ways in-vitro meat will change our lives. *h+Magazine*, 17 November. Retrieved on 30 April 2015 from http://hplusmagazine.com/2009/11/17/eight-ways-vitro-meat-will-change-our-lives/.

Knapp V (1997). The democratization of meat and protein in late eighteenth- and nineteeth-century Europe. *The Historian* 59(3): 541–51.

Krijnen M (2012). The need for meat: an interview with Mark Post. *Maastricht University Webmagazine*, 20 June. Retrieved on 30 April 2015 from http://webmagazine.maastrichtuniversity.nl/index.php/research/technology/item/271-the-need-for-meat.

MacDonald C (2009). Synthetic meat. *Biotech Ethics*, 15 June. Retrieved on 30 April 2015 from http://biotechethicsblog.com/author/ethicsblogger/page/9/.

Nimmo R (2011) Bovine mobilites and vital movements: flows of milk, mediation and animal agency. In J Bell (Ed). *Animal movements, moving animals: essays on direction, velocity and agency in humanimal encounters*. Centre for Gender Research, Uppsala, Sweden: Uppsala University Printers.

PETA (2011). PETA's 'in vitro' chicken contest. People for the Ethical Treatment of Animals. Retrieved on 30 April 2015 from http://www.peta.org/features/vitro-meat-contest/.

PETA (2015). Update to PETA's 'in vitro' chicken contest. People for the Ethical Treatment of Animals. Retrieved on 1 May 2015 from http://www.peta.org/features/vitro-meat-contest/.

Post M (2012). Cultured meat from stem cells: challenges and prospects. *Meat Science* 92(3): 297–301.

Rabinow P (1999). *French DNA: trouble in purgatory*. Chicago, IL: University of Chicago Press.

Reinert H (2007). The pertinence of sacrifice: some notes on Larry the luckiest lamb. *Borderlands e-journal* 6(3). Retrieved on 30 April 2015 from http://www.borderlands.net.au/vol6no3_2007/reinert_larry.htm.

Rose N (2007). Molecular biopolitics, somatic ethics, and the spirit of biocapital. *Social Theory and Health* 5(1): 3–29.

Szerszynski B (2010). Reading and writing the weather: climate technics and the moment of responsibility. *Theory, Culture & Society* 27(2-3): 9-30.

Stephens N (2010). In vitro meat: zombies on the menu? *SCRIPTed* 7(2): 394-401.

Twine R (2012). Revealing the 'animal-industrial complex': a concept & method for critical animal studies? *Journal for Critical Animal Studies* 10(1): 12-39.

Vialles N (1994). *Animal to edible*. Cambridge, UK: Cambridge University Press.

# 9

# Multispecies publics in the Anthropocene: from symbolic exchange to material-discursive intra-action

*Gwendolyn Blue*

In popular vernacular and academic scholarship, it is commonplace to assume that the public refers solely to humans. This assumption is understandable, given that the English term is derived from the Latin *publicus* 'of or pertaining to the people' and the Old French *public* 'open to general observation by the people'. Although an anthropocentric understanding of the public has been in play in modern society for hundreds of years, its days are arguably numbered. More-than-human organisms and technological objects are increasingly apparent in public life, leading scholars to reconfigure what counts as a political association as well as what it means to be human. These developments, far from being supplemental to human experience, have altered the very meaning and experience of 'life itself'.

Drawing conceptual resources from American pragmatism, through the work of John Dewey, and feminist science studies, through Karen Barad and Donna Haraway, this chapter examines the conceptual possibilities of positioning the public as a multispecies, rather than a strictly human accomplishment. This reformulation is motivated by

Blue G (2015). Multispecies publics in the Anthropocene: from symbolic exchange to material-discursive intra-action. In Human Animal Research Network Editorial Collective (Eds). *Animals in the Anthropocene: critical perspectives on non-human futures*. Sydney: Sydney University Press.

the complex entanglements constituting and constituted by the Anthropocene (Crutzen & Stoermer 2000, Steffen et al. 2007). The Anthropocene suggests that our collective ecological and political context has shifted, from the relatively predictable and benign climate of the Holocene in which humans developed as a species, to the unpredictable, chaotic, and increasingly dangerous environment in which the fruits of our collective labour have exposed the fragility of the planetary ecosystem where tipping points and planetary boundaries foil efforts to free ourselves from the constraints of natural systems (Latour 2010).

While the term is contested, the Anthropocene nevertheless promotes recognition of the intertwined social, biological, and geological elements of existence. It is an invitation, perhaps even an imperative, to extricate ourselves from the modernist conceit of a clean separation between humanity and the natural world, and between humans and other species. Its conceptual force lies with the acknowledgement that humans are interdependent with, and not separate from, the finite elements (air, water, soil) and geological timescales of a planet that can no longer be viewed as an inert backdrop on which the unfolding of human affairs takes place.

Moreover, as an increasing number of commentators suggest, the Anthropocene signals the emergence of a collective understanding of human subjectivity, a new 'we' that hangs together under the placeholder species (Chakrabarty 2009; Wilson 2009; Dibley 2012). To claim that humans have become a geological force means, as Dipesh Chakrabarty (2009) puts it, scaling up our imagination about what it means to be human. It is as a species, and not as individuals, that we have become a geological force. As a species, we are responsible for the present and continued loss of other species. Through land use changes, the spread of pathogens, the introduction of non-native species, the direct slaughter and indirect killing of animals and anthropogenic climate change, humans risk precipitating a mass extinction in the not-so-distant future (Barnosky et al. 2011). If the planet crosses certain ecological tipping points, we may also be endangering the existence of our own species.

Mired as it currently is in a humanist orientation, the public is an unlikely but important idea-companion, to borrow a term from Barad, for responding to the Anthropocene. Integral to Western democratic formations, the public provides a site of possibility for transformation and change. Politically speaking, it is important to recognise that the public is not a fixed term with stable meaning or referential context. This does not mean that we can define or can approach the public in any way we fancy. The concept has a history as well as an attendant scholarship that bears on how it is imagined and is enacted. Without an interrogation of the received wisdom about publics, we risk being inadequately responsive to the emergent entanglements and dynamic conditions within which new approaches to politics are emerging.

In what follows, a brief genealogy is offered in order to highlight the public's dynamic meaning, its normative appeal and its conditions of possibility. Two options are identified for thinking through the conceptual possibilities of a multispecies public. The first, focused on issues of identity, access and power, broadens the scope of the public to include non-human species. The second, focused on issues of communication and inquiry, draws on a processual view of publics as associations that lack an a priori identity or structure. From this view, publics emerge in circumstances that are continually born anew. It is the latter, elaborated through the work of John Dewey that forms the basis for a multispecies public. Next, an overview is provided of Karen Barad's notion of intra-action as a way of overcoming the anthropocentrism inherent in Dewey's account of the public. Drawing on Barad and Haraway, a multispecies public is defined as a processual phenomenon that, in turn, constitutes categorical distinctions and political divisions. The conclusion reflects on the implications of situating the public as a multispecies rather than a strictly human accomplishment.

## The public: a brief genealogy

While the Western usage of the term 'public' dates back to the ancient Greek notion of the agora as a meeting place for male, property-owning citizens, the modern public was initially established in tandem with democratic struggles against despotic states. Modern publics emerged in European salons, coffee houses and *Tischgesellschaften* (table societies) of Europe in the 17th and 18th centuries (Habermas 1989). In these spaces, philosophical and political ideas were discussed among emergent social classes that were markedly different from the ruling aristocratic patrons and professional elites of the time.

Through the course of the 19th century, a different inflection of the public developed with the professionalisation of science, an institution whose authority rested on its capacity to speak of, and for, the natural world (Broks 2006). The professionalisation of science required the presence of a public to witness, legitimate, support and finance its accounts of nature in the face of the established power of the Church. Science required the public and used it as a marker of its distinction. By the mid-20th century, in tandem with a mass-consumer society, buttressed in part by the professionalisation of journalism and broadcasting, science was established as a profession that spoke to, and not as, a public. As Latour (2005) highlights, this separation of science from its publics contributed to the modern sensibility that science speaks on behalf of the non-human by generating matters of fact whereas politics (and publics) represent people's interests and values, generating matters of concern.

Emerging from these historical configurations, there have been (at least) two overlapping but distinct scholarly approaches in the invention, refinement and popularisation of the public, and its associated concepts such as public sphere, public good and public life (Stob 2005). The first, and predominant, draws on issues of identity, access and power, raising questions about which individuals or groups are included in public life and which are excluded. Jürgen Habermas' (1989) bourgeois public sphere is an iconic example of this approach. Drawing on the conditions established in the formation of the modern public in Europe, Habermas puts forth a normative understanding of the public sphere as a space in which rational argument trumps social hierarchies, based on authority and status. Joan Landes (1988), Nancy Fraser (1992) and Jodi Dean (2002), among others, challenge Habermas' normative account of rational publics, arguing that the bourgeois public sphere privileges certain actors, discourses and practices while marginalising others. As Robert Asen and Daniel Brouwer (2001) argue, these critiques have transformed our understanding of the public sphere, directing attention to the interaction of multiple publics, to the political divisions that permeate society and to the inherently oppositional character of the citizenry in modern society.

An alternative, supplemental and complementary line of inquiry on publics focuses on the problems and possibilities of language in generating political communities and solving social problems. Here, the interest lies not with what the public *is*, in terms of the structures and identities that constitute the polity, but rather with what the public can *do* by constituting political communities through communication.

John Dewey's *The public and its problems* exemplifies this approach. Dewey argues against those who attribute political agency to independent, self-sufficient social actors. He also objects to the idea of a unified public that stands in distinction to a monolithic science. Dewey defines the public as 'all those who are affected by the indirect consequences of transaction, to such an extent that it is deemed necessary to have those consequences systematically cared for' (Dewey 1927, 16). For Dewey, the public does not have a pre-existing identity around which it adheres. Rather, publics emerge as a result of the interdependent nature of social relations that in tandem with the transformations wrought by science and technology, bring new problems into existence.

Dewey views all human actions and behaviour as emerging, not from self-contained individuals who are removed from the environments in which they are situated, but from ongoing transactions between organisms and environments. These transactions are a necessary, but insufficient, condition for the formation of publics. Inquiry and communication are also required in order for emergent publics to address the problems caused by the transactional nature of human existence. The ability to respond to the indirect consequences of interdependent action requires inquiry into and communication of the consequences that a set of actions has on others. The public's ultimate task, according to Dewey, is to confront emergent problems with an intimate understanding of the needs of the particular situation and to forge effective judgements that can inform intellectual and social practice in the context of a world that is ever changing, forever modifying and continuously in the making.

Dewey's transactional view of publics is predicated on the belief that humans differ from other animals and entities because of the human capacity for symbolic exchange:

human communities have traits so different from those which mark assemblies of electrons, unions of trees in forests, swarms of insects, herds of sheep and constellations of stars. When we consider the difference we at once come upon the fact that the consequences of conjoint action take on a new value when they are observed. For notice of the effects of connected action forces men [sic] to reflect upon the connection itself: it makes it an object of attention and interest. (Dewey 1927, 24)

In Dewey's account, although any entity can be enrolled in a public provided that the activities in question have consequences that affect it, only humans have the capacity to engage in inquiry and communication, and hence to constitute publics that are capable of responding to the issue at hand. Ultimately predicated on human exceptionalism, Dewey's approach to publics closes down the possibility that these associations might be more-than-human accomplishments.

## From transaction to intra-action

While Dewey's transactional view of publics offers a promising avenue for a multispecies public, his inherently humanistic perspective on inquiry and communication presents a significant barrier. Drawing on Barad's concept of intra-action, human-centred accounts of public life are decentred, replaced by an understanding of publics as phenomena that emerge as a result of the iterative material-discursive entanglements of a world that is continuously in flux.

Karen Barad (2007) coined the term 'intra-action' as a key concept in her agential realist account of the world. Intra-action is a foil to the metaphysics of individualism contained in, and by, the concept of interaction. Whereas interaction presupposes a world populated by discrete objects and subjects with essences that precede their engagement, intra-action stresses that actors do not engage as distinct, preformed entities, but rather are agents that mutually construct one another through ongoing relations. According to Barad, it is through intra-actions that articulations of the world become relevant through the constitution of boundaries and separations.

In developing her account of agential realism, Barad draws on the work of physicist–philosopher Niels Bohr, in particular, his recognition that we are embedded in the very realities we investigate. In Bohr's view, there is no definitive separation between investigating subjects and observed objects. Subject and object are not separate but constitute an integrated whole that Bohr refers to as a 'phenomena'. It is to these phenomena that observations refer, and not to objects that are independent of human perception. Reality is not separate from the manner by which it is observed, on the contrary, each object is given shape by virtue of the phenomena in which it is encountered.

While Bohr focuses his attention on laboratory apparatuses in making certain phenomena manifest and not others, Barad extends his insights to address everyday practices where 'apparatuses are to be understood not as mere laboratory instruments, static instrumental embodiments of human concepts, but as open-ended and dynamic material-discursive practices through which specific "concepts" and "things" are articulated' (Barad 2007, 334). In Barad's view, the world is comprised of phenomena made manifest as a result of ongoing material-discursive intra-activity. Here, a discursive practice is not the same as language, speech acts or symbolic action. It refers to 'specific material configuring of the world through which determinations of boundaries, properties, and meanings are differentially enacted' (Barad 2007, 335). Following Foucault and Butler, Barad defines discourse not in terms of what is said but as something that 'enables and constrains what can be said' (Barad 2007, 146). Discursive practices create the conditions in which meanings are made. Meaning is not solely a human achievement, but 'an ongoing performance of the world in its differential intelligibility' (Barad 2007, 335). In turn, matter is not external to discourse nor is it a fixed entity with a particular essence. Matter is a dynamic substance that is part of the ongoing intra-activity of the world.

Following Barad, publics can be understood as the material-discursive entanglements that emerge in tandem with technological, scientific, cultural and geological transformations. Publics do not precede their entanglement with the world – they emerge as a result of the material-discursive unfolding of the universe. The making of a public in any situation involves the creation of particular boundaries and distinctions that enable certain actions and meanings to emerge, while constraining others. Not all associations, however, are publics. To be public is to be rendered visible, a visibility that both requires and enables inquiry and communication. This inquiry must, in turn, be directed towards resolving the problem that brought forth the public in the first place. The point to be taken from Barad is that inquiry and communication are not strictly human achievements.

In referring to this particular formation of the public as a 'multi-species' achievement evokes what Haraway calls 'the ontics and antics of significant otherness, in the ongoing making of the partners through the making itself' (Haraway 2008, 165). In *When species meet* (2008), Haraway adopts Barad's concept of intra-action to describe the multispecies entanglements through which entities, organisms and bodies come into being. In line with Barad, Haraway envisions a world of bodies, not as self-contained autonomous entities, but as 'a multi-species crowd' (165), an entanglement of what she calls 'naturecultures' all the way down. Haraway maintains that we have never been (only) human, but that we are made and remade through complex entanglements, stretched across time and space, with other species. Our engagements with the world are comprised of intra-actions with diverse species. It is never a matter of 'us' removed from a world to which we turn our processes of inquiry and communication.

A multispecies approach to the public aligns with the broader project of feminist science studies that have long insisted on attending to the imbricated and mutual constitution of conceptual binaries (subject/object, nature/culture, human/non-human, science/society) and to how the constitutive exclusions that form as a result are enacted in practice. Rather than situating these conceptual divisions as a priori structures, feminist science studies directs attention to the 'cuts that bind', using Barad's terminology, taking into account the ways in which categorical distinctions emerge from, and as a result of, a world that is continuously in flux. For example, Lynda Birke, Mette Bryld and Nina Lykke (2004) use Barad's performativity to explore the ways in which the separation of animals from humans is enacted in particular settings, with a focus on the practices of science. Here, scientific practice is positioned as a discursive-material process that serves to constitute the very phenomena to which its inquiries are directed.

## Seeing ourselves as a species in the Anthropocene

How can a multispecies public help us think through the emergent configurations put in play by the Anthropocene?

As discussed earlier, the concept of species has emerged as a way of envisioning human subjectivity in an era in which the boundaries between the human and natural world have been breached. In the foreward to economist Jeffrey Sach's *Common wealth*, a book that also uses the idea of the human species in the Anthropocene, Edward Wilson remarks that:

> humanity has consumed or transformed enough of Earth's irreplaceable resources to be in better shape than ever before. We are smart enough and now, one hopes, well informed enough to achieve self-understanding as a unified species . . . we could be wise to look on ourselves as a species. (Wilson 2009, xii)

In doing so, Wilson claims that the sciences should play a more central role. As he writes in a commentary in the *New York Times* (2013):

> the task of understanding humanity is too important and too daunting to leave to the humanities . . . they have not explained why we possess our special nature and not some other form of a vast number of conceivable possibilities. In that sense, the humanities have not accounted for a full understanding of our species' existence. (Wilson 2013)

A multispecies public is an intervention into this emergent species discourse. A multispecies public is not a call for the self-mastery of humans as a collective; it calls into question the constitutive exclusions on which fantasies of unification are based. While inquiry and communication play a central role in its foundation, a multispecies public does not trade in the modernist division between authoritative facts and a gregarious populace. Science matters in the formation of multispecies publics, to be sure, but not as an authoritative, disembodied voice that hides its situated contingency under a cloak of universality. The boundaries, limits and even the potential of the Anthropocene are neither mute fact nor political fiction. They are enactments of an emergent public of which science is a part that is responding to the unintended consequences of collective action.

## Conclusion

Drawing on a Deweyian understanding of publics as entities formed in, by, and through, everyday transactions, multispecies publics are positioned, not as a formation with a discernible structure to which questions of identity, access and power can be applied in a straightforward fashion but as a dynamic process that unfolds and transforms in an ever-changing interdependent world. As Dewey highlights, interdependency is a necessary but insufficient condition for the constitution of a public. Communication and inquiry are also required. Barad and Haraway, in turn, offer conceptual resources for considering how communication and inquiry can be characterised not by virtue of strictly human-driven symbolic exchanges but by the enmeshment of material-discursive practices that give rise to meaning. It is not a question of who ultimately creates this meaning because the capacity to generate meaning is a relational matter, born, as Haraway puts it, of the 'material-semiotic requirements of getting on together in specific life-worlds' (Haraway 2008, 263).

The point of a multispecies public is not to widen the bounds of the public sphere to include everyone and everything, human and non-human, animate and inanimate. On the contrary, a multispecies public seeks to decentre the human political subject, resonating with the broader ethical project of human–animal studies and feminist science studies. These fields of inquiry have long insisted on attending to the mutual constitution of conceptual binaries and the ways in which these binaries promote and enact exclusions. Ultimately, multispecies publics raise questions about response-ability in the context of the transformations brought forth by the Anthropocene. As Barad emphasises, responsibility is not shaped by a predeterminative set of conditions. Neither calculation nor intention, the capacity to respond flows out of the cuts that bind, emerging from the ongoing intra-active becomings that constitute the world.

The Anthropocene can be thought of as a collective experiment in public making, one that is simultaneously material and discursive. It is contributing to new visions of political subjectivity with emergent forms of inclusion and exclusion that warrant attention. Contained in, and by, the parameters of inquiry and communication, its formulations are fundamental to the ways in which we live together, presently and in the future. The point we wish to make is that the 'we' in question is not simply one species among many. Not only are we not separate from the world to which we direct our inquiry, we are not, and have never been, only human.

## Works cited

Asen R & Brouwer D (2001). *Counterpublics and the state.* Albany: SUNY Press.

Barad K (2007). *Meeting the universe halfway: quantum physics and the entanglement of matter and meaning.* Durham, NC: Duke University Press.

Barnosky A, Matzke N, Tomiya S, Wogan G, Swartz B, Quental T, Marshall C, McGuire J, Lindsey E, Kaitlin C, Maguire B & Ferrer E (2011). Has the Earth's sixth mass extinction already arrived? *Nature* 471(7336): 51–57.

Birke L, Bryld M & Lykke N (2004). Animal performances: an exploration of the intersections between feminist science studies and studies of human/animal relationships. *Feminist Theory* 5(2): 167–83.

Broks P (2006). *Understanding popular science: issues in cultural and media studies.* Berkshire, UK: Open University Press.

Chakrabarty D (2009). The climate of history: four theses. *Critical Inquiry* 35(2): 197–222.

Crutzen P & Stoermer EF (2000). The 'Anthropocene'. *Global Change Newsletter* 41: 17–18. Retrieved on 29 April 2015 from http://www.igbp.net/download/18.316f18321323470177580001401/NL41.pdf.

Dean J (2002). *Publicity's secret: how technoculture capitalizes on democracy.* Ithaca, NY: Cornell University Press.

Dewey J (1927). *The public and its problems.* Athens, OH: Ohio University Press.

Dibley B (2012). 'Nature is us': the Anthropocene and species-being. *Transformations* 21. Available online at: www.transformationsjournal.org/journal/issue_21/article_07.shtml.

Fraser N (1992). Rethinking the public sphere: a contribution to the critique of actually existing democracy. In C Calhoun (Eds). *Habermas and the public sphere* (pp109–42). Cambridge, MA: MIT Press.

Habermas J (1989). *The structural transformation of the public sphere: an inquiry into a category of bourgeois society.* T Burger (Trans). Cambridge, MA: MIT Press.

Haraway D (2008). *When species meet.* Minneapolis, MN: University of Minnesota Press.

Latour B (2005). From Realpolitik to Dingpolitik – or how to make things public. In B Latour & P Weibel (Eds). *Making things public: atmospheres of democracy* (pp4–31). Boston, MA: MIT Press.

Latour B (2010). A plea for earthly sciences. In J Burnett, S Jeffers & G Thomas (Eds). *New social connections: sociology's subjects and objects* (pp72–84). Basingstoke, UK & New York: Palgrave Macmillan.

Landes J (1988). *Women and the public sphere in the age of the French Revolution.* Ithaca, NY: Cornell University Press.

Steffen W, Crutzen P & McNeill J (2007). The Anthropocene: are humans now overwhelming the great forces of nature? *Ambio* 36(8): 614–21.

Stob P (2005). Kenneth Burke, John Dewey and the pursuit of the public. *Philosophy and Rhetoric* 38(3): 226–47.

Wilson EO (2009). Foreword. In J Sachs. *Common wealth* (ppvi–xiii). New York: Penguin Books.

Wilson EO (2013). The riddle of the human species. *New York Times*, 24 February. Retrieved on 30 April 2015 from http://opinionator.blogs.nytimes.com/2013/02/24/the-riddle-of-the-human-species/.

# 10

# Apiculture in the Anthropocene: between posthumanism and critical animal studies

*Richie Nimmo*

The concept of the Anthropocene encapsulates an interesting paradox; it contains the seeds of divergent ways of thinking which, if not quite contradictory, certainly pull in different directions. This chapter explores the implications by tracing this tension through the relationship between critical and posthumanist currents in human–animal studies. The theoretical discussion is worked through and grounded in a case study of a contemporary environmental crisis involving non-human animals, specifically the escalating crisis of honeybee apiculture that has come to be known as colony collapse disorder (CCD).

At one level the notion of the Anthropocene could be regarded as quintessentially posthumanist in its firm rejection of the notions of human separation from – and transcendence of – the natural world that are at the core of the modern humanist worldview. Instead the Anthropocene asserts the thoroughly terrestrial nature of human beings as *Homo sapiens*, and emphasises our interconnectedness with – and inescapable dependence on – the planetary biosphere. But there is also within the concept the lingering root of a fundamental humanist idea.

Nimmo R (2015). Apiculture in the Anthropocene: between posthumanism and critical animal studies. In Human Animal Research Network Editorial Collective (Eds). *Animals in the Anthropocene: critical perspectives on non-human futures*. Sydney: Sydney University Press.

An acknowledgement of the sheer scale of the effects of human activity upon the planet is at the heart of the concept and underpins the force of its invitation to critical self-reflection; but in accomplishing this by stressing that no other animal species has ever had a comparable impact on the Earth, it simultaneously invites the sort of emphasis on the ostensibly exceptional nature of human beings that features so prominently in humanist thinking. Though intended as a warning and a call for change, this slips all too easily into the characteristic hubris of humanism, inadvertently reaffirming its secular theology of unlimited human self-creation, autonomy and self-mastery. Thus the Anthropocene concept contains what one might think of as an ambivalent posthumanism, inviting us to reflect on our existential status as one biological species among many others inhabiting a finite planet, while seeming to continue to whisper the humanist promise that we are still the centre of the world and capable of being the absolute masters of our fate. The former is a call for human modesty about being human; the latter is anything but that.

Yet something resembling this lingering humanism seems difficult to avoid within any discourse that calls for collectively taking responsibility, for action and for change, not least at the level of the species. A certain residual humanism is entangled here with the adoption of a 'critical' approach, since a recurrent feature of critique – or more precisely, the mode of critique commonly mobilised within forms of ecocriticism – is that it seeks not only to describe a state of affairs, and perhaps to change ways of thinking about reality, but also marks a call to action and a demand for structural change in the world; the point, as Marx famously argued, is to change it. This sort of critical stance has attracted criticism for its tendency to suppose that an external vantage point on the object of critique is somehow attainable, and thence to elevate the ostensibly privileged knowledge of the critical theorist above that of everyday knowledge (Boltanski 2011; Cooper 2013, 83–85). Extending this, one might make the related point that this kind of critical strategy is predicated on – indeed, insists on – the attribution of ultimately determinative responsibility to human beings, which in turn, presupposes a rather strong conception of human agency. If we are to believe that we can use our critical understanding of the world in order to change it decisively, we then have also to believe that we possess the capacity to do so, which is to say that we have the ability not only to transcend our circumstances and seize our destiny but to do so collectively as a species. As Boria Sax points out with devastating honesty, this is the archetypical humanist conceit:

> Each of us may individually feel nearly helpless as we contemplate a vast number of threats from personal bankruptcy to global warming ... Nevertheless, 'humanity' itself seems nearly omnipotent. Since it is rather like the way that the Greeks personified lightning as Zeus and the sea as Poseidon, the personification of humankind as a dominant figure is, in very literal ways, a myth. Far from being unified, we human beings barely keep our tendency toward mass slaughter of one another under fragile and sporadic control. We are nowhere remotely close to being able to consciously guide the course of history or even the evolution of technology. 'Anthropocentrism' is this tendency to vastly exaggerate human dominance, understanding, power, autonomy, unity, guilt, virtue, wickedness, and morality. (Sax 2011, 35–36)

There is, therefore, perhaps not quite an irreconcilable contradiction, but certainly a persistent tension, between critical approaches underpinned by a humanist ontology and rooted in a politics of responsibility and transformation, and the sort of posthumanism that would play down the decisiveness of human agency and stress instead that, insofar as it is a significant term at all, agency is not an exclusive property of human beings – in fact it is not a property of entities, but must be understood as relational.

In this vision, agency is not just about the interplay of reflexive human intentions and social–structural conditions, as in most sociological accounts, but is emergent from hybrid assemblages encompassing relations between heterogeneous actants, both human and non-human (Latour 1993; 2005, 63, 70, 75–76; Michael 2000, 1). In the posthumanist refiguring of agency developed in Karen Barad's 'agential realism', for example:

> Agency is a matter of intra-acting; it is an enactment, not something that someone or something has. Agency cannot be designated as an

attribute of subjects or objects (as they do not preexist as such) . . . it seems not only appropriate but important to consider agency as distributed over nonhuman as well as human forms. (2007, 214)

Thus agency is shaped by the co-presence of multiple intra-acting material-discursive and biosocial networks, such that human intentions are merely one element in an open-ended and dynamic ensemble of more-than-social relations. Consequently, human intentions are variously generated, frustrated, mediated, transformed and enabled by other entities and by other energies, such that contingencies proliferate and unintended consequences are legion; as Haraway puts it '[h]istorical specificity and contingent mutability rule all the way down, into natureculture, into naturecultures' (2003, 300). In this way posthumanism decentres the humanist subject-agent by unmasking the co-constitutive entanglement of humans and non-humans in a materially heterogeneous world (Knappett & Malafouris 2008). Such refusal to be anthropocentric about agency, in turn, complexifies and problematises the residually humanist notions of responsibility that tend to be implicit in many 'critical' approaches, pointing instead towards a different kind of responsibility and a different sort of politics.

## Colony collapse disorder

The phrase 'colony collapse disorder' was first used early in 2007 to refer to the alarming phenomenon of dramatic, large-scale and unexplained disappearances of honeybee colonies, initially among commercial pollinators in Florida and California during the autumn of 2006 (Cox-Foster, Frazier et al. 2007). These were believed to be distinct from the periodic losses that are a routine hazard of beekeeping due to a number of characteristics: firstly the sheer scale and suddenness of the losses – with large and apparently thriving colonies sometimes disappearing in no more than a fortnight; also due to the peculiar absence of the normal invaders, such as wax moths or beetles, that would usually take advantage of the absent colony to consume the significant stores of honey remaining in the hive. Another distinctive feature was the relative absence of the bee carcasses that would normally be found within the hive and littered around the entrance to collapsed colonies. Moreover, those carcasses that were discovered, when later subjected to pathological analysis, were found to be suffering from a bewildering array of viruses and infections, as though their immune systems had collapsed. These features combined to create the sense that this was something never before encountered; 'colony collapse disorder' – or CCD – was born. After some initial scepticism, the scale of colony losses across 22 states of the USA by the spring of 2007 had given credence to the idea that CCD was real, and it began making headlines worldwide in the context of raising alarm at the potential impact on pollination, hence food production and the agricultural economy, if such a rapid rate of colony losses continued. Before long reports of honeybee losses across the world were being linked to CCD, in Canada, Taiwan, Spain, Portugal, Italy, France, Poland, Austria, Belgium, the Netherlands, Croatia, and the UK.

The first scientific research programs attempting to identify the cause of CCD began soon after. The plausible candidates included: the widespread use of the new class of pesticides known as 'neonicotinoids', developed in the 1980s and widely used since the 1990s, nerve toxins that were suspected of detrimental effect on the apian nervous system, even at sub-lethal doses; the apparently unstoppable progress of parasites such as the parasitic *Varroa* mite and associated viruses such as the *Nosema* infection and Israeli Acute Paralysis Virus, introduced into vulnerable honeybee populations by migratory pollination and the under-regulated transcontinental trade in honeybees; a loss of genetic diversity due to poor breeding practices favoured by some commercial breeders; the possible unintended consequences of GM crops in weakening bee immune systems; the over-intensive exploitation of honeybees in monocultural commercial pollination, leading to intolerable migratory stresses on that species and crowding out native pollinators; the increasing frequency of unseasonal weather, known to affect the reproductive and foraging cycles of bee colonies, due to accelerating climate change; and, finally, changing landscapes involving the loss of areas of diverse flora such as wildflower meadows that long played a vital role in sustaining native bee populations.

Despite ongoing investigations at numerous institutions, none of these has yet emerged as a convincing candidate for a sole causal explanation of CCD. Early claims by entomologists and genetics analysts to have isolated the pathogenic agent most directly responsible for CCD as Israeli Acute Paralysis Virus proved premature, and were soon discredited, but not before imported honeybees from Australia were erroneously identified as the source of the problem in the USA, with significant consequences for Australian bee exporters (Cox-Foster, Conlan et al. 2007; Anderson & East 2008). Much subsequent research has stressed the major roles played in the ongoing problem of honeybee decline by *Varroa, Nosema* and other pathogens, but CCD cannot be convincingly attributed to any one of these alone (Genersch et al. 2010). Moving from biological pathogens to human-made toxins, there are sharply conflicting views on the effects of neonicotinoid pesticides on bees, with both sides able to selectively cite entomological and pathological research that appears to support their position (Maxim & van der Sluijs 2013; Schmuck 2013; Henry et al. 2012). Many beekeepers, some of whom have found their livelihoods threatened by CCD, have been convinced for some time – often based on several decades of experience – that neonicotinoid pesticides are centrally to blame, and many have campaigned vigorously with a range of environmental groups to have those chemicals banned. They secured a significant victory in April 2013 when the European Union imposed a moratorium on the use of neonicotinoids for two years in member-states pending further investigation, on the basis of a report by the European Food Safety Authority (EFSA) which concluded that these pesticides pose an 'acute risk' to honeybees (EU Regulation 485/2013). Meanwhile pesticide manufacturers such as Bayer Crop Science and Syngenta, whose profits rely on routine agricultural use of these products, have lobbied intensively against this, and have funded research on the causes of CCD intended to help exonerate neonicotinoids. In September 2013 these companies began legal action against the EFSA in an attempt to overturn the ban. Clearly then, the debate about CCD has been far more than simply an arena of disinterested scientific investigation; it has been a cosmopolitical struggle between different ways of knowing 'nature', contrasting ontologies, and rival visions of society–nature relations.

With CCD having been first recognised in 2007, it seems increasingly unlikely that any specific causal agent will be identified as being responsible, and the scientific consensus belatedly emerging is that the phenomenon may be multi-causal, with several factors potentially interacting in highly complex ways (vanEngelsdorp et al. 2009; Neumann & Carreck 2010, 1–6; Williams et al. 2010, 846). It nevertheless remains unknown whether the acceleration of honeybee decline marked by CCD represents the intrusion of some new, and as yet, unverified pathogenic or toxic agent into the honeybee ecosystem, as was initially suspected and as the anti-pesticide campaigners maintain, or if it is better understood as a ratcheting-up of the many known pressures on honeybees. The following section takes this up by beginning to delineate the basis of a critical animal studies approach to understanding CCD.

## Honeybees as livestock

Media references to CCD often invoke 'a crisis of bees'. There is such a crisis, consisting of a long-term serious decline of native pollinators, but this is not synonymous with CCD, which is more precisely a crisis of the Western honeybee, *Apis mellifera*. This is an agricultural–economic crisis as much as an environmental one, since the Western honeybee is the pollinator of choice for much of the world's commercial agricultural crops (Klein et al. 2007, 303–13). In developing a critical perspective on apiculture it is useful to consider categorisations of honeybees as a kind of 'lilliputian livestock – fuzzy herbivores with wings' (Buchmann & Nabhan 1997). The species has been selected for its prolific rate of honey production and pollination, transported around the world with global flows of colonial power and capital, and is now heavily relied on as an intrinsic component of the system of agricultural mass production known as monoculture (Williams et al. 2010, 845). This can be seen most clearly in the almond industry in central California, where 644 km of orchards depend entirely on honeybees trucked in seasonally by commercial pollination operators from across the USA – and when necessary flown in from Australia – in order to pollinate its 60 million almond trees, which supply more than three quarters of the world's almonds (Traynor 1993; Singeli 2007). This is not an isolated case; Californian almond farming is the most dramatic example of an intensively industrialised monoculture, heavily reliant on commercial apiculture for pollination, but it is by no means the only example, and it represents the model towards which many other branches of agriculture are moving, driven by the relentless logic of rationalisation, standardisation and intensification. As with many other forms of animal agriculture, this process is most advanced in the USA, but the particularity of the American case should not be overstated; large-scale monocultures around the world are similarly dependent upon the 'pollination services' provided by *Apis mellifera*, the most prolific of all pollinators (Aizen & Harder 2009, 915–18).

Clearly then, insofar as CCD is characterised by the sudden and large-scale collapse of honeybee colonies, it marks a serious crisis of agriculture, much of which would quickly be rendered non-viable if honeybees were to disappear (Aizen & Harder 2009, 915–18). A sobering glimpse of this scenario can be observed in Sichuan Province in China where every pear tree has had to be painstakingly pollinated by human workers since all honeybees were wiped out by pesticide misuse more than 20 years ago (Tang et al. 2003, 14–15, 18). This case has only underlined the value of *Apis mellifera*, as the best efforts of human pollination have been unable to even approximate the efficiency of honeybees and have resulted in much lower levels of fertility, and at a huge financial cost. To imagine this scenario repeated throughout world agriculture is chilling in its implications. To regard CCD as a 'crisis of bees', however, is to take the dominance of *Apis mellifera* as given and to marginalise the many other species of native bees that performed the work of pollination perfectly well before the globalisation of the Western honeybee and the industrialisation of agriculture (Buchmann & Nabhan 1997). It also normalises the monoculture system of food production with its increasing dependence on commercial honeybee apiculture for pollination. According to this way of thinking, a crisis of the honeybee is a world crisis of natural fertility – an almost apocalyptic scenario from which it follows that the only solution is to 'save the honeybee' in order to return to business as usual. A critical approach, in contrast, would involve refusing to normalise the status quo, pointing to the contingency and unsustainability of the current system, and stressing the urgent need for a less intensive and more diverse agriculture.

In this light, the irony of the ubiquitous 'save the honeybee' sentiments and campaigns that have sprung up since 2007, however well meaning, is that the international proliferation of the honeybee is inseparable from the forms of agricultural organisation that have led to the steady decline of native pollinators and have engendered the rationalising practices that ensured it was only a matter of time before something like CCD occurred. As scholars in critical environmental sociology have long argued, the natural world is not as infinitely malleable as modernism likes to believe; complex natural systems and processes are amenable to only so much human manipulation and intensification before they tend to reach some sort of critical tipping point, giving rise to any number of unforeseen consequences (Catton & Dunlap 1980; Benton 1989; 1993; Dickens 2004, 100–103, 115). This may be something of a well-worn realist refrain, but it is borne out by a litany of environmental crises in recent decades, from bovine spongiform encephalopathy (BSE) to avian flu, and from antibiotic resistance to climate change. In short, therefore, a system based so exclusively upon the pollination work of a single species is manifestly unsustainable, and from a critical perspective it is this, rather than any single specific causal agent, that is ultimately behind the honeybee's decline. That is not to deny the likely role of pesticides such as neonicotinoids in significantly accelerating CCD, but a critical approach means going beyond the search for mono-causal explanations and acknowledging that a genuinely sustainable solution would have to involve not only a far less chemical-intensive agriculture but also a managed decline of industrialised apiculture and a diversification of pollination.

A critical approach to apiculture then is valuable on a number of fronts: firstly, drawing on the suggestive concept of the 'animal-industrial complex' developed by scholars in critical animal studies (Noske 1989, 22–39; Adams 1997; Twine 2012), its structural lens provides a means to grasp what we might call the 'apis-industrial complex'. This avoids focusing more narrowly upon the ethical and environmental problems pertaining to practices that constitute fragmentary elements of this complex, such as pesticide use or long-distance migratory pollination. Considered in isolation these may appear to be potentially solvable through reform, within the status-quo, but a critical approach demands acknowledgement that all such problems ultimately stem from the industrial capitalist political-economic structure of the system. As David Nibert (2003, 17) puts it, 'such practices are inevitable in a selfish and profit-driven economic system that has fostered agricultural concentration'. This, in turn, lays the ground for arguing consistently for a structural transformation and radical alternative (Best 2009). In addition, by grasping honeybees in terms of their structural role and position within this complex, hence as a kind of 'livestock', a critical approach is well positioned to explain why this species – like many other intensively farmed 'livestock' animals – is being pushed to its biological limits, and very likely beyond, by processes of rationalisation and intensification. In this way a critical approach effectively grounds a politics of opposition to commercial apiculture as an instance of a much wider industrial–animal agriculture or 'agribusiness'.

## Of hives and hybrids

Yet this is not quite satisfactory empirically, as there are significant differences between honeybees and the animals usually signified by the term 'livestock'. The various species of mammals that make up the majority of agricultural animals are far more unambiguously 'domesticated' than honeybees, with a long cultural history of being discursively positioned in some ambivalent conceptual space between 'wild' and 'domesticated'.

According to the humanist ontological architecture underpinning the concept, domestication involves a species becoming no longer 'wild', as human intervention, manipulation and control move the species from the sphere of 'nature' to that of 'culture'. Even for those such as Stephen Budiansky (1999) and Roger Caras (2002) who want to stress the agency of non-human animals in the process of domestication, so that it is no longer synonymous with domination and is seen as having been in some sense 'chosen', the result of this co-evolutionary process is still that the animal is no longer 'wild'. In this way of thinking, kept honeybees are domesticated insofar as beekeepers routinely make interventions into their reproductive processes, enabling them to manipulate the rhythms of the colony, aiming, for example, to inhibit the occurrence of colony division through swarming and thus promote uninterrupted honey production (Chandler 2009, 91). Kept honeybees also live in human-made hives specially constructed in such a way as to facilitate the extraction of honey by humans without destroying the honeycomb and damaging the hive. In these respects, kept honeybees are comparable to other 'livestock'.

Yet honeybee colonies manifest such complex social organisation and specialised division of labour that honeybees have long been seen as social beings in their own right, not just as animals that have been domesticated by virtue of being enfolded within the structure of a human 'social' domain, or, conversely, as part of a wild nature defined by its separation from human society. This wild/domesticated binary relies on an anthropocentric equation of sociality with humanity, and non-humanity or nature with a-sociality, such that to be social is to have been humanised and thereby denaturalised; but bees problematise this by asserting their existence as social creatures, with their own complex form of sociality. Honeybee colonies were perceived as micro 'societies' many centuries before developments in ethology began to attribute sociality to a range of non-human mammals such as primates, and they continue to be regarded as the paradigmatic 'social insect', inviting myriad conceptualisations of the 'politics' of the hive (Preston 2006; Seeley 2010). Some of the most recurring symbols of collectivity and sociality in numerous cultures are apian or bee-related, and an oft-repeated beekeeping proverb states that *una apis, nulla apis* – 'one bee is no bee', underlining the significance of the fact that apiculture is always the relationship of a human individual or group of humans with a collective comprising many thousands of bees. Beekeepers do not enter into relationships with individual bees, they do not concern themselves with the health or welfare – let alone the 'rights' – of any individual bee; every individual is a manifestation of a multitude, the colony, and it is this multitude with which beekeepers are engaged. Hence 'almost none of the standard western ideas of individuality and autonomy of self have any purchase in the study of bees' (Preston 2006, 15).

It is also significant that the activities of honeybees have long been seen as a form of 'work', as labour in an almost human sense, not simply as natural behaviour. This is reflected in Marx's famous comparison of the skill and ingenuity of honeybees, in the construction of their honeycomb chambers, with that of human architects (1976, 284). He goes on to argue that the difference between the labour of bees and that of human architects is that, whereas the bees operate on instinct, according to a blueprint drafted by evolution, the human architect designs the structure consciously in the mind before constructing it in reality (Ingold 1983; Benton 1993). In this way Marx toys with the non-anthropocentric idea of bees as labourers, hence as social beings, before moving to place bees firmly back within the domain of nature, by contrasting their ostensible lack of conscious agency with that of human beings.

It is worth carefully noting the conceptual status of labour that underpins such thinking. Labour is conceived of as purposeful activity upon nature, transforming something natural – a raw material – into an artefact or manufactured good, into something useful, a social thing. Marx refers to honeycomb, but it is honey itself that is most telling in this respect. Honey is commonly seen as a naturally occurring raw material and, at the same time, as in some sense manufactured, because it is produced within highly organised apian societies by 'worker' bees dedicated to its production. Claude Lévi-Strauss, for example, located honeybees in a liminal zone between nature and culture, pointing out that even wild honeybees are markedly 'civilised' in their collective and organised labour of transforming nectar into honey, and suggesting that honey is thus better seen not as 'raw' but as 'cooked' (1973, 28, 35, 55, 289). Yet the cooks here are non-human, and in this sense honey is still categorised as 'natural' because, apart from the work of managing the hives and periodically extracting the ready-made honey and putting it into jars, no human labour enters into its production; the alchemic work of converting nectar into honey is performed entirely by the bees. This is reflected in the marketing of honey as a 'natural' product, and often as a 'natural' alternative to sugar and other 'artificial' sweeteners. Thus the ambiguous cultural status of honey as both given by nature and manufactured, raw and cooked, natural and cultural, underlines the highly liminal status of honeybees themselves, as non-humans engaged in highly organised production of an artefact valued by humans. Other animals produce things of value to humans, but not through such highly organised collective activity that lends itself so easily to being categorised as 'work'; and other animals produce things through highly organised activity, but not things that are highly valued by human beings. Moreover, there are no other 'livestock' animals whose products we consume, but whose bodies we do not.

For these reasons, beekeeping has tended to be perceived and to be represented not simply as another example of domestication but as a more hybrid or mutual endeavour, involving a form of interspecies collaboration between bee societies and human societies. Certainly honeybees are not easily categorised as domesticated agricultural animals without neglecting much of what is specific and unique about human–apian relations. As Claire Preston puts it, '[i]f bees have some of the instincts of the herd, they are not precisely *domestic* animals. Fortunately, they consent to inhabit artificial hives which have been devised for them, but their relationship to man is better conceived as symbiotic, with each species benefiting from certain behaviours and capabilities of the other' (2006, 34). Honeybees are not subjected to the sort of human confinement and close control that arguably defines the lives of 'cattle'; the modern beehive is not the equivalent of a fenced-in field or cage, let alone a factory farm; it is not a disciplinary container bounded by the human will-to-control, but is better described as an apparatus that stages and mediates a co-constitutive interaction between bodies, an iterative material enactment the outcome of which is always at least partly open-ended, contingent, and negotiated.

Beekeepers have traditionally described their activities in terms that are distinct from those of the farmer, or those of the petkeeper, but which are instead suggestive of a practice with its own specific set of ontological coordinates on the cultural map of human–animal relations (Wilson 2005, 231–71). In the words of one practising beekeeper, 'the way forward is to work closely *with* the bees, developing a relationship based on mutual benefit and cooperation rather than simple exploitation' (Chandler 2009, 7). Such practitioners' understandings should perhaps not be taken at face value, but nor can they be dismissed out of hand in a presumption of superior insight on the part of the critical theorist. The role is that of a bee *keeper* and the activity is bee *keeping*, not bee farming or bee minding, and this terminology is not insignificant. It is consistent with the term api*culture*, where the etymological root of 'culture' refers to cultivation – one *cultivates* bees, hoping to create the conditions for them to flourish and to produce excess honey. However, one cannot simply force them to do so through the imposition of ever-greater control, in the way that cattle may be forced to provide milk, chickens to provide eggs, and all animals to provide meat. If too much honey is removed from the hive and there is not enough left for the colony then it will not survive the winter, which is a serious loss and failure for the beekeeper (Preston 2006, 35–36). Thus, unlike some other forms of animal agriculture, increased productivity in this case depends on the animals thriving and cannot easily be accomplished instrumentally at a cost to their wellbeing, which renders it problematic to regard honeybees as 'dominated' or 'exploited'.

## From domination to trust

For all the strengths of a critical animal studies approach, then, understanding honeybee apiculture by means of a critical analysis of the exploitative human–animal relations constitutive of the category of 'livestock', means that many of the distinctive characteristics of bees and beekeeping are brushed aside or left unacknowledged for the sake of critical consistency. Posthumanism provides an important corrective, by insisting that we think through the lived specificities of beekeeping practices and the nuances of human–apian relations in their irreducible materiality. This, in turn, makes it possible to grasp apiculture non-anthropocentrically as a hybrid human–non-human assemblage, and to explore the ways in which beekeeping practices and discourses, as well as the mode of existence of bees, may engender ways of knowing and being that confound the nature/culture and human/non-human binaries that underpin both the conceptual architecture of 'domestication' and its 'critical' critique.

In an influential essay, Tim Ingold presents an 'indigenous' account of domestication in which, rather than a narrative centring on the passage of animals from an original state of wildness, defined by a lack of human control, to a state of domestication, defined by the imposition of human control, instead the central transition is from human–animal relations based on trust to relations based on domination (1994, 18). Relations of trust are here defined as a state in which something is freely given to another in the hope that it will be reciprocated, but with no certainty that it will be, and without any element of compulsion (Ingold

1994, 16). An important implication is that the other party retains genuine agency in the relationship, since they may reciprocate, but are not forced to do so. Ingold argues that this characterises the relations between hunter-gatherers and the animals they hunt, since their belief is that by treating the hunted animals with respect, eschewing cruelty and unnecessary infliction of pain, and avoiding wasting the animal once killed, the hunter maintains good relations with the species in question. This is essential to ensure that future hunts have a fair chance of success because the animals are believed to have the power either to present themselves to the hunter, thus enabling him to eat, or to refuse to appear, so that the hunter will go hungry (Ingold 1994, 13, 14). In this way the animals are granted significant – even decisive – agency in the relationship, at least in terms of the hunters' definition of the situation. Ingold contrasts this with the human–animal relations that predominate in farming where, however affectionately the farmer may think of the animals, and however well they may be treated, they are understood as essentially subservient and dependant; they are not believed to have the power to significantly affect the farmer's fortunes, or to withhold what the farmer wants from them. Farm animals are regarded as being under almost complete human control, and as lacking agency; hence no trust is required on the part of the farmer, since this is essentially a relationship of domination (Ingold 1994, 17).

Reflecting on this, it is striking just how significant a role is played by trust in beekeeping. By making exhaustive preparations and taking all of the right precautions, the beekeeper hopes that his or her bees will be more likely to thrive, to produce abundant honey and to survive the winter; but there is no certainty of this, and even the most experienced beekeepers will have encountered disappointment or disaster, usually more than once, and sometimes inexplicably. The complexity of honeybees, their colonies and their finely calibrated interrelationship with the local environment, means that there are always many contingencies in play, rendering notions of complete control alien to beekeeping. As one practising beekeeper puts it, '[t]he bees know what they are doing: our job is to listen to them and provide the optimum conditions for their well-being' (Chandler 2009, 36). In this sense, beekeepers are more like Ingold's hunter-gatherers than his farmers; they do what they can, observe the correct rituals, and hope that their diligence will be repaid; but they do not have or seek the power to ensure that the desired outcome is achieved, or to force bees to produce honey, and they will not speak of their activities in these terms. Beekeepers tend to have an acute awareness that they are dealing with living and dynamic complexity and must remain open to contingency, and they will not deny the agency of the bees in the enterprise. As is evident, for example, in this particularly articulate but not untypical reflection on an early formative experience of beekeeping:

> standing there in my beekeeper's suit, lording it over the hives, I could say I was assuming a stance that ignored the clear rules of the bees' nature. It didn't feel like arrogance in the moment. I suppose hubris rarely does. But certainly I was overestimating my own competence and abilities ... I had an *idea* of myself as a beekeeper and was acting accordingly. I was getting stung and bees were dying because my idea of myself as a beekeeper was getting in the way of seeing what was really going on. (Magill 2010, 10)

Pursuing a parallel posthumanist thread, Donna Haraway's ontology of 'companion species' foregrounds 'co-constitution, finitude, impurity and complexity' (2003, 302). In this multispecies vision, it is not just that non-humans and humans 'possess' agency, but that agency itself is conceived as relational and distributed, something perpetually emergent from heterogeneous 'actor-networks' – to borrow Bruno Latour's term (1993; 2005) – rather than somehow preceding them. Thus 'none of the actors precede, finished, their interaction. They more than change each other; they co-constitute each other, at least partly' (Haraway 2003, 307); companion species are constitutively and materially entangled and entwined together within the forever unfolding biosocial web of life. There is no getting outside of this, and the human subject, far from being transcendent over nature, is relationally co-constituted vis-à-vis non-humans in an ongoing process of material-semiotic becoming. Haraway's examples are drawn from 'dog worlds' and the intricately entangled histories of human–canine co-evolution, but this ontology could well have been dreamt up by a meditating beekeeper, so uncannily do honeybees fit the companion species mould. In these terms, it is not a question of the relations between separate sovereign

entities, 'humans' and 'bees', but of the iterative enactment of human–bee hybrids, the simian–apian assemblages that have proliferated
since *Homo sapiens* first came into contact with *Apis mellifera*, tasted
its honey, felt its sting, observed its indefatigable activity, and began to
imagine the possibilities.

## Entangled politics in the Anthropocene

Returning to CCD and the crisis of apiculture, a posthumanist politics
might proceed by engaging closely with the lived human–apian entanglements that constitute beekeeping, tracing the many symmetrical and
heterogeneous elements of this deeply historical and hybrid practice.
Instead of treating beekeepers as little more than intensive livestock
farmers, they might be regarded as interspecies practitioners, senso-
rially and materially engaged in a liminal world interceding between
the purified categories of 'nature' and 'culture'. Beekeeping becomes not
just an unsustainable and exploitative practice per se, but a contested
terrain, subject to rationalising forces without question, but also potentially replete with relations of trust, forms of decentred interspecies
encounter, and moments of affective connection with vital materiality, non-human being, and more-than-human value. Thus the task of a
posthumanist critique is to relentlessly excavate and to foreground the
cosmopolitics of 'living with' others that are incipient within apicultural
practices, and to advocate for forms of practice and organisation that
nurture these entangled ways of knowing and being. Rather than beginning by asking what is wrong here; what is unacceptable; and what
must stop; the key questions become what is good here; what is of value;
and how can this be developed. In this way, understanding the world of
beekeeping as a hybrid multispecies assemblage can underpin a more
nuanced and constructive politics of beekeeping.

In terms of concrete political positions there may well be significant overlap with the substantive politics of critical animal studies, and
there is nothing to be gained by artificially magnifying the differences.
To an extent the two approaches may be taken to refer to different
objects, and insofar as this is so they may not be in direct conflict
and could even be rendered complementary. For example, one might

suggest that while a posthumanist understanding of small-scale or amateur beekeeping is apt, this breaks down somewhat when considering very large-scale apiculture and commercial pollination operations, with their ecologically unsustainable monocultures, economies of scale and intensive profit-driven practices. Conversely it could be argued that while the critical–structural approach is highly persuasive as an analysis of big commercial pollination enterprises, it is stretching credulity to try to apply this sort of structural critique to small-scale or hobby beekeepers. Such a pragmatic position is perhaps preferable to a rigid theoretical purism that would insist on the universal applicability of any single way of thinking about the world. But it would surely be a mistake to conclude that posthumanist analysis can therefore safely be applied to more 'acceptable' human–animal practices while more troubling human–animal relations must be reserved for the critical animal studies treatment. Such out-and-out pragmatism would not only still mean glossing over the specificities of honeybees, apiculture and human–apian relations in order to assimilate large-scale beekeeping to a critical–ethical framework developed with reference to intensively farmed mammals, it would also elide the real differences between these ontologies and their performative politics of the 'human'.

Crucially, a genuinely posthumanist politics is never just about seeking to transform human relations with non-human animals, however important this may be; it is always also about seeking ways to simultaneously transform our most fundamental relations with ourselves as human, changing how we see and experience ourselves and our relationship with the world – our mode of existence, our very way of being human. It is a vision fuelled by humility as much as ethical conviction, and by a sense of modesty about humanity, rather than righteousness; this will tend to lead to a different kind of interspecies engagement and a different sort of politics – at times messier, more tentative, more willing to trust perhaps, and more tolerant of apparent contradiction. As an explorative sensibility orientated towards an entangled collective future that is always unfolding and yet to be fully understood, posthumanism can certainly be exuberant, but it will not presume to have all the answers already worked out, grounded in ethical universals, and just waiting to be implemented by an overriding human agency.

What can be learnt from this about the implications for human–animal studies of the paradoxical posthumanism embedded in the concept of the Anthropocene, and how to respond to it? The Anthropocene and the predicament it names poses more sharply than ever the dilemma now facing humanity in a time of unprecedented ecological crisis. In underlining the profound impact of human activity on the planetary biosphere and the geosphere itself, the Anthropocene should drive the final nail into the coffin of humanist notions of human autonomy and separation from nature. In the Anthropocene humans are finally unmasked as earthlings, as *Homo sapiens*, far from the transcendent beings of the techno-humanist imagination, unbound by material constraints; this dethroning of the human subject is consistent with a posthumanist sensibility. It is double-edged, however, because insofar as the Anthropocene is intended or is interpreted as a didactic concept and not simply a descriptive one, it powerfully reaffirms human agency and human responsibility by implying that we must take control of the planetary consequences of our species' activity in order to shape our own fate; a deeply humanist sentiment. In this respect the Anthropocene discourse contradicts itself: pointing to the organic interconnection of humanity and the natural world, as evidenced by our impact upon the Earth and the coming consequences of this for human society; while demanding that we rise above nature by taking control of our collective species activity and consciously modifying it, in what would be the ultimate demonstration of humanist transcendence and self-mastery.

As first this appears intractable but, as this discussion has aimed to show, tracing the same paradox into the tension between posthumanist and critical approaches in human–animal studies allows the problem to be approached somewhat differently. Thus, applied to CCD the Anthropocene underlines the unsustainability of current levels and forms of human productive activity, specifically industrial commercial apiculture, and its severe detrimental impact upon natural systems. It also emphasises human dependence upon these natural systems, and predicts the eventual negative, possibly catastrophic, consequences for human beings of their breakdown. Indeed CCD is exactly the sort of crisis that we should anticipate facing with increasing frequency if we are living in the Anthropocene. When it comes to the question of what is to be done though, things are a little more complicated. The critical reading of the Anthropocene would stress the need for large-scale structural change in apiculture, amounting to either the abolition or significant diminution of commercial pollination and honey production. But a posthumanist reading would suggest that this is not consistent with a proper understanding of the situation the Anthropocene diagnoses.

By stressing the profound interdependence of humanity and the natural world, now realised on a geospheric level, the Anthropocene effectively acknowledges that humanity and nature are so inextricably intertwined that any attempt to move forward by re-establishing a more ethical separation and properly regulated interaction between the two domains is surely misconceived; there are no such separate domains. Paralleling this, the 'critical' politics of human responsibility and transformation cannot adequately address the problems of hybrid socio–natures in which agency is distributed between multiple heterogeneous actants and responsibility is at best partial and fragmentary. The humanist dream of human sovereignty, transcendence and self-knowing was always part of the anthropocentric mode of existence that impelled the ecological crisis at the heart of the Anthropocene; this crisis will not be solved by urging that the same humanist dream be realised at a still higher level. Hence a consistently posthumanist politics cannot be framed as a call for social–ecological transformation tantamount to a transcendent act of human responsibility at a species level; instead it must try to develop a more explorative, local and entangled politics of finding more humble ways to live together with others in the hybrid and finite colony we share.

## Works cited

Adams CJ (1997). 'Mad cow' disease and the animal-industrial complex: an eco-feminist analysis. *Organization and Environment* 10(1): 26–51.

Aizen MA & Harder LD (2009). The global stock of domesticated honeybees is growing slower than agricultural demand for pollination. *Current Biology* 19(11): 915–18.

Anderson D & East IJ (2008). The latest buzz about colony collapse disorder. *Science* 319(5864): 724–25

Barad K (2007). *Meeting the universe halfway: quantum physics and the entanglement of matter and meaning.* Durham, NC: Duke University Press.

Benton T (1989). Marxism and natural limits: an ecological critique and reconstruction. *New Left Review* 178: 51–86.

Benton T (1993). *Natural relations: ecology, animal rights and social justice.* London: Verso.

Best S (2009). The rise of critical animal studies: putting theory into action and animal liberation into higher education. *Journal for Critical Animal Studies* 7(1): 9–53.

Boltanski L (2011). *On critique.* Cambridge, UK: Polity Press.

Buchmann S & Nabhan G (1997). *The forgotten pollinators.* Washington, DC: Island Press.

Budiansky S (1999). *The covenant of the wild: why animals chose domestication.* New Haven, CT: Yale University Press.

Caras R (2002). *A perfect harmony: the intertwining lives of animals and humans.* West Lafayette, IN: Purdue University Press.

Catton WR & Dunlap R (1980). A new ecological paradigm for a post-exuberant sociology. *American Behavioral Scientist* 24(1): 15–47.

Chandler PJ (2009). *The barefoot beekeeper.* PJ Chandler, http://www.biobees.com/.

Cooper G (2013). A disciplinary matter: critical sociology, academic governance and interdisciplinarity. *Sociology* 47(1): 82–85.

Cox-Foster D, Conlan S, Holmes EC, Palacios G, Evans JD, Moran NA, Quan PL, Briese T, Hornig M, Geiser DM, Martinson V, vanEngelsdorp D, Kalkstein AL, Drysdale A, Hui J, Zhai J, Cui L, Hutchison SK, Simons JF, Egholm M, Pettis JS & Lipkin WI (2007). A metagenomic survey of microbes in honey bee colony collapse disorder. *Science* 318(5848): 283–87.

Cox-Foster D, Frazier M, Ostiguy N, vanEngelsdorp D & Hayes J (2007). 'Colony collapse disorder': investigations into the causes of sudden and alarming colony losses experienced by beekeepers in the fall of 2006. Colony Collapse Disorder Working Group. Pennsylvania State University.

Dickens P (2004). *Society and nature: changing our environment, changing ourselves.* Cambridge, UK: Polity Press.

European Union Regulation 485/2013, 24 May 2013. Amending Implementing Regulation (EU) No 540/2011, as regards the conditions of approval of the active substances clothianidin, thiamethoxam and imidacloprid, and prohibiting the use and sale of seeds treated with plant protection products containing those active substances. *Official Journal of the European Union*, L139: 12–26.

Genersch E, Ohe W, Kaatz H, Schroeder A, Otten C, Büchler R, Berg S, Ritter W, Mühlen W, Gisder S, Meixner M, Liebig G & Rosenkranz P (2010). The German bee-monitoring project: a long term study to understand periodically high winter losses of honey bee colonies. *Apidologie* 41(3): 332–52.

Haraway D (2003). Cyborgs to companion species. In *The Haraway reader* (pp295–320). New York: Routledge.

Henry M, Béguin M, Requier F, Rollin O, Odoux JF, Aupinel P, Aptel J, Tchamitchian S & Decourtye A (2012). A common pesticide decreases foraging success and survival in honeybees. *Science* 336(6079): 348–50.

Ingold T (1983). The architect and the bee: reflections on the work of animals and men. *Man: Journal of the Royal Anthropological Institute of Great Britain and Ireland* 18(1): 1–20.

Ingold T (1994). From trust to domination: an alternative history of human–animal relations. In A Manning & J Serpell (Eds) *Animals and human society: changing perspectives* (pp1–22). London: Routledge.

Klein AM, Vaissière BE & Cane JH (2007). Importance of pollinators in changing landscapes for world crops. *Proceedings of the Royal Society of Biological Sciences* 274(1608): 303–13.

Knappett C & Malafouris L (2008). *Material agency: towards a non-anthropocentric approach*. New York: Springer.

Latour B (1993). *We have never been modern*. Harvard, MA: Harvard University Press.

Latour (2005). *Reassembling the social: an introduction to actor-network theory*. Oxford, UK: Oxford University Press.

Levi-Strauss C (1973). *From honey to ashes: introduction to a science of mythology, II*. London: Cape.

Magill M (2010). *Meditation and the art of beekeeping*. Lewes, UK: Leaping Hare Press.

Marx K (1976). *Capital: a critique of political economy*. Vol. I. London: Penguin.

Maxim L & van der Sluijs J (2013). Seed-dressing systemic insecticides and honeybees. In *European Environment Agency Report No.1/2013. Late lessons from early warnings: science, precaution, innovation* (pp369–406). European Environment Agency.

Michael M (2000). *Reconnecting culture, technology and nature: from society to heterogeneity*. London: Routledge.

Neumann P & Carreck NL (2010). Honey bee colony losses. *Journal of Apicultural Research* 49(1): 1–6.

Nibert D (2003). Humans and other animals: sociology's moral and intellectual challenge. *International Journal of Sociology and Social Policy* 23(3): 5–25.

Noske B (1989). *Humans and other animals*. London: Pluto Press.

Preston C (2006). *Bee*. London: Reaktion Books.

Sax B (2011). What is this quintessence of dust? The concept of the 'human' and its origins. In R Boddice (Ed). *Anthropocentrism: humans, animals, environments* (pp21–36). Leiden, The Netherlands: Brill.

Schmuck R (2013). The Bayer CropScience view on Maxim and van der Sluijs' 'Seed dressing insecticides and honeybees'. In *European Environment Agency Report No.1/2013. Late lessons from early warnings: science, precaution, innovation* (pp401–2). Copenhagen, Denmark: European Environment Agency.

Seeley TD (2010). *Honeybee democracy*. Princeton, NJ: Princeton University Press.

Singeli A (2007). The almond and the bee. *San Francisco Chronicle*. SFGate.com, 12 October. Retrieved on 30 April 2015 from http://www.sfgate.com/magazine/article/The-Almond-and-the-Bee-2518870.php.

Tang Y, Xie JS & Chen K (2003). Hand pollination of pears and its implications for biodiversity conservation and environmental protection: a case study from Hanyuan County, Sichuan Province, China. Sichuan University College of the Environment.

Traynor J (1993). *Almond pollination handbook*. Bakersfield, CA: Kovak Books.

Twine R (2012). Revealing the 'animal-industrial complex': a concept and method for critical animal studies? *Journal for Critical Animal Studies* 10(1): 12–39.

vanEngelsdorp D, Evans JD, Saegerman C, Mullin C, Haubruge E, Nguyen BK, Frazier M, Frazier J, Cox-Foster D, Chen Y, Underwood R, Tarpy D & Pettis JS (2009). Colony collapse disorder: a descriptive study. *PLoS ONE* 4(8): e6481.

Williams GR, Tarpy DR, vanEnglesdorp D, Chauzat MP, Cox-Foster DL, Delaplane KS, Neumann P, Pettis JS, Rogers REL & Shutler S (2010). Colony collapse disorder in context. *Bioessays* 32(10): 845–46.

Wilson B (2005). *The hive*. London: John Murray.

# 11

# The welfare episteme: street dog biopolitics in the Anthropocene

*Krithika Srinivasan*

On 4 March 2007, the municipal corporation of Bengaluru, India, launched a drive to remove street dogs from the city. The drive, called 'Operation Dog Hunt' by the media, was triggered by public outcry over the mauling and death of a four-year-old boy a few days previously. This attack was preceded by a similar incident that left an eight-year-old girl dead earlier that year. More than 1000 dogs were rounded up in the days that followed, and many were killed. By 8 March 2007, there was public outcry again, though this time, against the rounding up and killing of the dogs. These protests brought Operation Dog Hunt to a halt fewer than five days after its launch (*Deccan Herald* 2007a).

The reaction to Operation Dog Hunt is but one instance of publicly articulated concern for non-human life that is increasingly seen in the contemporary world (Franklin 1999; Murdoch 2003; Youatt 2008). While such concern is most obviously seen in domains of public activism and political debate such as environmentalism and animal welfare, questions about how to live ethically with non-human life forms

Srinivasan K (2015). The welfare episteme: street dog biopolitics in the Anthropocene. In Human Animal Research Network Editorial Collective (Eds). *Animals in the Anthropocene: critical perspectives on non-human futures*. Sydney: Sydney University Press.

are taken up by a much wider range of actors, and at varied scales – individual, institutional, national and global (Hobson 2007).

Another notable characteristic of these times is their christening as the Anthropocene (Crutzen & Stoermer 2000) in recognition of the scale and consequences of human impacts on the planet and the various life forms which live on it. The growing body of scholarship – in the physical, natural and social sciences and the humanities – on the Anthropocene has mainly paid attention to 'the emergence of human action as a critical force in a range of biophysical systems' (Kotchen & Young 2007, 149). In doing this, the literature exhorts humankind to take responsibility for, and respond to, the harm that has been associated with their activities (Johnson et al. 2014). What has received less attention in this literature is the concomitant rise of global human concern and action for planetary, ecological and animal wellbeing. As discussed above, the Anthropocene is characterised not just by harmful and exploitative human activity but also by widespread and fairly strong manifestations of human responsibility and care for non-human Others.

It is these contradictory features of the Anthropocene that are at the core of this chapter. In essence, the chapter explores the complexities that arise when these two opposing trajectories come together, that is, when harmful human ways of life intersect with efforts to ethically improve interactions with non-human Others. Drawing on empirical research on street dog welfare in India and on Michel Foucault's writings (1977, 2002, 2008) on epistemes and the flows of power in spaces of care and reform, I examine the domain of animal welfare to offer a critical account of the manners in which animal wellbeing comes to be conceptualised and pursued in contexts circumscribed by normative objectives related to not only animal wellbeing but also human wellbeing. In essence, by using debates around street dog welfare in Bengaluru to think about relations of care where humans seek to protect the wellbeing of animals even while safeguarding human interests, my goal here is to not only question the self-evidence of what are accepted as discourses and practices of straightforward improvement but also trace the unexpected ways in which power can infuse human–animal interactions in the Anthropocene.

## Street dogs in India

The domestic dog is found in most human cultures and locations across the world (McHugh 2004). In most developed countries, dogs are seen as animals that can live only under human care – whether in homes, shelters, laboratories or farms. 'Stray' dogs in these countries are almost always perceived as abandoned or escaped pets – as 'out-of-place' (Philo & Wilbert 2000). India, on the other hand, like many other countries such as South Africa, Romania, and Russia, has free-living dogs that thrive in both rural and urban areas (Duijzings 2011). These dogs are not merely out-of-place pets; neither are they feral animals that fear human beings (Jackman & Rowan 2007, 57). Their relationships with people are varied: Indian street dogs can be seen as pests or nuisances; can be tolerated or ignored; can be cherished, fed and sheltered as neighbours or companions (Srinivasan 2013; Nolan 2006).

While street dogs are an integral part of the cityscape in Bengaluru, for more than a century, local authorities here (as in other parts of India and the world) have sought to eliminate them for reasons of human health, with rabies being a particular concern. Before 2001, street dog control programs in India (introduced initially during British colonial rule) involved killing by electrocution, bludgeoning, poisoning or shooting (Krishna 2010). In 2001, the passing of the Animal Birth Control (Dogs) Rules made it mandatory for all local authorities seeking to control dog populations to adopt the trap-neuter-vaccinate-release strategy. Typically, these animal birth control (ABC) and anti-rabies vaccination (ARV) programs are implemented in collaboration with animal welfare organisations.

The ABC/ARV program has not been uncontroversial. Questions about the place of free-living dogs in India made national and international headlines in 2007 when two children were killed by street dogs in Bengaluru (Beary 2007). The protests that followed provide a window to the recurrent debates about street dogs and how humans should (or should not) live alongside them. In Bengaluru, it was initially argued that street dogs should be eliminated by means of 'euthanasia' because of the threats of biting, mauling and rabies. There was also concern for India's image. As a member of the group Stray Dogs Free Bangalore emphasised, 'in any civilised country, you don't see dogs wandering around'.[1] Stray dog control, then, was also a matter of aesthetics, of national pride and of attaining the status of 'developed country' (*Times of India* 2007). Such ideas of development are predicated on industrial modernisation, identified as the key driver of the changes that have led to the labelling of these times as the Anthropocene (Crutzen & Stoermer 2000).

The Operation Dog Hunt, cobbled together by the municipal authority in an attempt to appease public anger, equally evoked opposition, as outlined in the introduction. Animal activists and members of the public protested against the cruel treatment and killing of dogs during Operation Dog Hunt, instead advocating neutering, vaccination, and domestic waste management[2] as more effective and humane (Compassion Unlimited Plus Action WP 427 2007; *Deccan Herald* 2007b; WP 4920 2007). Operation Dog Hunt was stopped after five days, but these heated discussions and intensive ABC/ARV programs continued for the next few months.

At first glance these arguments about human–dog interactions in Bengaluru exhibit fairly irreconcilable positions about how humans ought to relate to dogs. However, a closer examination reveals 'regularities' (Foucault 1991a) in the form of consensus on broad normative objectives on both human and animal wellbeing. For one, the sights and sounds of the dogs being caught and killed during Operation Dog Hunt horrified large sections of the public. There was general agreement that humans should not cause 'unnecessary suffering' to street dogs (MoEF WP 427 2007, 21). Interviews with those who were and are strong advocates of street dog eradication (including the group Stray Dogs Free Bangalore) indicated that even they did not want the dogs to be treated badly: 'when they came to catch during the drive, it was indeed cruel . . . Anyone who sees it will say it is cruel. We need a scientific way of luring the dogs away to some other place and then sending[3] them to where they have to go'.[4]

---

1 Interview, Stray Dogs Free Bangalore, 2010.
2 In order to reduce easy availability of food for street dogs.
3 'Sending' was used as a euphemism for killing.
4 Interview, senior government official, 2010.

On the other hand, there was, including among the animal activists, unanimity that street dog populations must be reduced and managed in order to meet human interests: 'there should not be too many dogs . . . Nobody supports the presence of too many dogs.'[5] This second point of normative consensus, that is, consensus about human wellbeing, is of significance here. While social science literature (for example, Agrawal 2005; Mol et al. 2010) has variously theorised the universalisation of norms about non-human interests/wellbeing, this chapter reflects on the entrenchment of norms about the sanctity of human interests, and the consequences for the manners in which animal wellbeing is conceptualised and is pursued.

## The welfare episteme

It is useful to draw on Foucault's writings on epistemes to think about these debates and the normative consensus about both human and animal interests. While discussing academic legitimacy, Foucault points to regularities that operate within and across disciplines and theorises these as episteme (2002). To Foucault, the episteme is that which 'makes it possible to grasp the set of constraints and limitations which, at a given moment, are imposed on discourse' (2002, 211). The episteme refers to 'the taken-for-granted assumptions of a regime' (Legg 2005, 147), and forms the 'space of dispersion' (Foucault 1991b, 55) for thought and action, defining the limits and possibilities of the latter at particular points in time.

As discussed, the debates about street dogs exhibit not only a dispersion of opposing discourses but also regularities that cut across them. Following Foucault, these regularities can be understood in terms of an episteme – which I call the welfare episteme – that enables and legitimises certain kinds of discourses and practices, and not others, in the context of human–street dog interactions.

While episteme refers to the purely discursive, the term apparatus or *dispositif* refers to a set of discursive and more-than-discursive practices that are particular to a social domain (for instance, the penal

5   Interview, Animal welfare practitioner (AWP), 2010.

system), and that operate within and are co-constitutive with an episteme (Foucault 1980, 194–97). In this chapter, I refer to the ensemble of truth discourses and material practices directed at fostering street dog wellbeing as the street dog welfare dispositif, and go on to show how it functions within the welfare episteme.

The welfare episteme has the twin normative objectives of safeguarding human interests – health, convenience and aesthetics – even while ensuring that the dogs do not undergo 'undue' suffering. In the debates examined, the epistemic norms require that street dogs must be managed in order to protect human health, safety, convenience and aesthetics. But dog wellbeing is equally an important concern. As an infectious disease specialist explained, most people:

> want a humane approach to be adopted toward the dog. Nobody wants a defenceless animal to be killed, you know, and dogs are endearing creatures. But in the face of this kind of attack – the death of children – how do you balance the two? (Interview with an infectious disease specialist and anti-rabies campaigner, Bengaluru, India 2010)

The balancing act that the above quote refers to is a reflection of the contradictory goals of the welfare episteme: even though there is overarching consensus about the importance of both human and animal wellbeing, these are normative objectives that are in tension because human and dog wellbeing are not directly correlated here. The presence of street dogs does pose risks to human health and safety and can offend human aesthetic sensibilities; similarly, insulating humans from risks and inconveniences posed by street dogs would require the complete elimination of street dog populations.

The normative boundaries delineated by the welfare episteme are not unique to the case at hand and can be seen in animal welfare practice in general. Animal welfare[6] is now seen as the 'moral orthodoxy' (Garner 2004, 10); it gives form and voice to the idea that 'it is morally wrong to inflict "unnecessary harm" on animals or treat them in ways

---

6  Animal welfare can be distinguished from animal liberation and animal rights as distinct modes of animal protection (Munro 2012).

that are not considered "humane" ' (Francione & Garner 2010, ix). The concept of 'unnecessary harm' as a founding principle of animal welfare reflects the balancing act between human and animal interests that animal welfare sets out to achieve. The aim is to manage situations so that existing ways of relating to animals can continue, but in a manner that is considered 'humane' (Schmidt 2011). Animal welfare discourse is thus often permeated with win-win approaches – win-win governmentalities (Dean 2010) – where both human and animal interests are simultaneously addressed (Twine 2010; Cole 2011).

Epistemes influence how individual humans think and act in particular domains. However, the analytical value of this concept lies in its description of assumptions or values that operate at a scale that is more-than-individual. The values and choices embedded in an episteme are so normalised that they tend to be seen as facts, and any attempt to question them or to ask for clarification appears absurd, and at times, even offensive or just plain wrong. While such values or choices or assumptions might be questioned by individuals and/or in private, such questioning is typically considered unacceptable, especially in the public/societal spheres (Igoe et al. 2010, 505). Epistemes incorporate assumptions or values at the scale of the 'rule' rather than the 'exception' at a particular moment, in the context of a particular social domain. As Foucault (2002, 69–70) explains, these are rules that 'operate not only in the mind or consciousness of individuals, but in discourse itself; they operate therefore, according to a sort of uniform anonymity, on all individuals who undertake to speak in this discursive field'.

Such rules can be identified in the welfare episteme. For instance, it is accepted that 'unnecessary suffering' (in animals) is bad. Similarly, there is no overt questioning of why human interests are important, for example, why human deaths due to rabies or dog bites are 'bad', or why it is 'bad' when humans are offended aesthetically or inconvenienced by these animals. It is worth noting that just the act of listing these human interests and questioning their moral value can appear ludicrous.

## Intervening on and for dogs

The explicit identification of the welfare episteme is useful because it directs attention to the complexities that emerge when spaces of care such as street dog welfare are shaped by competing normative objectives. My core argument is that the twin epistemic objectives of dog and human wellbeing have the effect of rendering biopolitical the pursuit of street dog welfare. For this, I look to Foucault's theories of the subtle and diverse ways in which power functions in society and penetrates all layers of social interaction. In particular, I draw on his work on forms and mechanisms of power that are not obviously harmful, that is, non-sovereign modes of power such as disciplinary and biopolitical power that seek to govern and to foster the flourishing of life at the level of the population (Foucault 1977, 2003, 2008).

A core characteristic of disciplinary and biopolitical power is that even when they work on and through the bodies of individuals, it is the population or the collective which is the main subject–object of these modalities of power. Related to this is an entanglement of harm and care that comes about because the individual is often used as a means to an end in the pursuit of the welfare of the larger population or collective (Foucault 2008; 2009, 42–44). Biopolitical power is also marked by the eschewal of sovereign mechanisms that forbid or repress; by contrast, the exercise of biopower involves governmental techniques that work alongside existing rhythms in the biosocial collectivity[7] in order to achieve win-win situations (Foucault 2009, 352; Dean 2010, 44). I show below how the dog welfare dispositif displays several features that indicate the flow of non-benign biopower in this space of care and reform.

Street dog welfare in Bengaluru has been addressed mainly through the animal birth control and anti-rabies vaccination program (ABC/ARV). This program, while administered by the municipal authority, is implemented by animal welfare organisations (AWO) and forms a core part of their animal welfare activities – and of the activities

---

7   The term 'biosocial collectivity' is an extension of Foucault's term 'population' to include more-than-human groupings of humans and animals (Holloway & Morris 2012).

of AWOs across the world (Srinivasan 2013). The ABC/ARV program, by neutering dogs and vaccinating them against rabies, aims to maintain stable, reduced and safe populations of street dogs in the city. This intervention is seen as a win-win approach as it has 'the potential to improve *both human and animal welfare* and avoid ethical conflicts' (Chief Health Officer WP 427 2007, my italics).

Neutering and vaccination work with natural rhythms in canine behaviours and biological processes. Vaccination is a classic biopolitical technique as it is based on the 'rationalisation of chance and probabilities', and works alongside and seeks the support of the disease[8] to prevent it by creating a reservoir of vaccinated and therefore 'safe' dogs (Foucault 2009, 59). Neutering takes into account and uses other existing realities. Dogs, being territorial animals, prevent the entry of new dogs into their neighbourhoods. Complete removal of dogs from a neighbourhood invites the entry of new, possibly diseased and aggressive, dogs from other areas, as long as food and space is available. The ABC/ARV program retains vaccinated and neutered dogs in their original neighbourhoods, preventing the influx of new dogs or the 'compensatory breeding of dogs to fill ecological niches' (Jackman & Rowan 2007, 65).

The interventions of neutering and vaccination also display the biopolitical characteristic of being directed at dogs as populations. This is even though it is particular individual dogs that killed the two children, and particular individual dogs that bite or transmit rabies. And yet, in the 2007 debates, there was no mention of the specific dogs that did the mauling. It remains uncertain what happened to those dogs. The focus was instead entirely on managing and on regulating the population as a whole, rather than dealing with the individual dogs linked to the incidents.

When it comes to the objective of dog wellbeing as well, the focus is on the management of the population. Interviews with animal welfare practitioners in Bengaluru (and abroad) reveal that the ABC/ARV program is seen as enabling dog welfare because a 'high-quality' population – one that is neutered, small, vaccinated, non-aggressive, healthy – contributes to overall dog wellbeing by warding off negative human

---

8   The rabies virus is used to build protection against the disease.

attention. Furthermore, it is argued, by animal welfare practitioners both in Bengaluru and elsewhere that 'overpopulation itself is a welfare problem for dogs' because of competition for food sources and the spread of infectious diseases (Jackman & Rowan 2007, 59). Neutering is also perceived as promoting dog wellbeing by foreclosing future dog lives that can potentially suffer – as one animal welfare practitioner said, 'instances [sic] such as pups getting crushed under vehicles reduce, if there are fewer pups'. Neutering as an animal welfare practice works by ensuring the wellbeing of dogs as a population. In totality, neutering is a biopolitical intervention in which certain behaviours and biological processes of the individual members of a population are identified as threatening 'the general welfare and life of the population' and thereby become the target of efforts 'to confine, to contain, to coerce, to eliminate, if only by prevention' (Dean 2010, 119, 171).

On the whole, the street dog welfare dispositif avoids sovereign interventions whether the electrocution or poisoning of all dogs or granting all dogs the right to autonomous lives. Instead, it incorporates typically biopolitical interventions that function by 'finding support in the reality of the phenomenon, and instead of trying to prevent it, making other elements of reality function in relation to it, in such a way that the phenomenon is cancelled out, as it were' (Foucault 2009, 59). In other words, the ABC/ARV program works with and uses dog bodies and behaviours, and manages dogs as a population in the pursuit of solutions of care that are in conformity with the win-win objectives of welfare episteme, that is, so that human and dog interests are simultaneously addressed.

### Individuals and collectives, harm and care

Such win-win management is not, in and of itself, problematic. However, the overarching approach of intervening in dog bodies to meet the twin goals of the welfare episteme is associated with an intermingling of harm and care that is a classic attribute of biopower. Even though biopower is directed at fostering life, this does not mean that the violence and harm disappear; rather, they are rationalised as necessary for the flourishing of the population, of the collective (Foucault 2008,

137). The individual often becomes a means to an end in the exercise of biopolitical interventions that are concerned with 'the processes that sustain or retard the optimisation of the life of a population' (Foucault 2009, 42; Dean 2010, 119).

As I have discussed elsewhere (Srinivasan 2013), the intervention of neutering dogs illustrates how animal welfare practices can be harmful and are yet perceived as protective because of the focus on dogs as a population. The innocuous word 'neutering' refers to castration in males and ovariohysterectomies in females. Scientific and activist literatures mostly describe only the benefits of neutering such as the prevention of uterine infections and certain types of cancers (ovarian, testicular, mammary), reduction in wandering behaviours and aggression, and associated increases in longevity (Jackman & Rowan 2007, 66). In general, the possibility that neutering might have adverse health impacts is overlooked or is downplayed[9] in both scientific and animal protection circles (Palmer et al. 2012).

Nonetheless, it has been observed at least by some that apart from the stress, pain, trauma and other side effects (including death due to surgical complications) caused by castrations and ovariohysterectomies (Palmer et al. 2012), neutering impacts individual canine health in various ways, right from causing incontinence, weight gain and associated diseases to increasing the risks of some cancers (Reichler et al. 2003; Liptak et al. 2004; Elliot 2008; ASPCA 2012). On the whole, the 2012 review of veterinary literature suggests that neutering (on otherwise healthy animals) does not enhance, and can even adversely affect, the wellbeing of individual dogs (Palmer et al. 2012). Medical issues are additional to the inherently problematic nature of the surgical curtailment of reproductive behaviours, something that the welfare episteme does not usually consider. Forced castration and sterilisation would not be considered acceptable in the human context and poses not dissimilar ethical questions when it comes to other animals.

The harmful aspects of 'coercive sterilisation' (Palmer et al. 2012, S163) are further exacerbated by the processes of capture, sheltering and release. The capture of dogs for the ABC/ARV program can be very

---

9   My experiences in animal welfare practice in both India and the UK corroborate this.

traumatic, as can the kennelling (Nolan 2006; Jackman & Rowan 2007). My personal observations in the course of my extensive involvement in the animal birth control program in India confirm this.

Capture is usually with catchpoles or lassos and involves rough handling. Shelter conditions are crowded and unhygienic; nosocomial infections such as canine distemper, pneumonia, parvovirus, and kennel cough are recurring problems, and often fatal. Common post-operative complications include infections of the scrotum in male dogs and the infection and opening of sutures in female dogs. It is also not unheard of for dogs to come out of the influence of the anaesthetic during surgery. Female dogs are sometimes found after release with open incisions, and sometimes, exhibiting evisceration of internal organs through the incision wound. As Nolan (2006) notes, these programs 'if done badly . . . may almost be as harmful as the alternatives [electrocution, poisoning etc.]'.

It is important to note that these are problems that are almost unavoidable given the large numbers of dogs handled and limited financial, material and human resources. Animal welfare practitioners (AWPs) are not oblivious of these complexities. One remarked, 'I think in many ways we have been very cruel to the dogs, the kinds of things we do to them, it is not funny at all. I wouldn't like it done to me.'[10] Another practitioner even suggests that neutering is not necessary for reasons of human health and safety as the restricted availability of food and habitat acts as a ready-made check on dog population growth: 'nature will find a way of controlling the population'[11] (also see Jackman & Rowan 2007, 72).

Despite these many serious harms associated with neutering (Palmer et al. 2012, S164), it is advocated as an intervention of care, and as an animal welfare measure rather than merely an animal control strategy. As discussed earlier, AWPs view it as fostering the welfare of dogs as a population. Neutering thus intervenes harmfully on individuals to protect the wellbeing of the dog population (and the human–animal biosocial collectivity) as a whole: care and harm are knotted together in this profoundly biopolitical act of improvement.

---

10  Interview, AWP, 2010.
11  Interview, AWP, 2010.

## Immersion and subjectification

The chapter has so far laid out a set of arguments that collectively indicate that the dog welfare dispositif, in seeking to address the competing normative demands associated with the welfare episteme, is a space marked by the operation of non-benign biopolitical power. In the context of human interactions, biopower is underpinned by processes of subjectification wherein the subject-objects of power self-govern, that is, act upon themselves, sometimes with harmful consequences, in accordance to truth discourses (Rabinow & Rose 2006; Foucault 2008). Subjectification makes efficient the operation of power by making questions about the positive, neutral or harmful impacts of self-governance unthinkable.

When it comes to dogs (and other animals) however, the non-human subject-objects do not self-govern: dogs do not offer themselves up for neutering on the basis of human-produced discourses about the benefits of castration and ovariohysterectomies. Rather, biopower in these more-than-human interactions is facilitated by agential subjectification (Srinivasan 2013; 2014) wherein human actors (AWPs, in this case), act for, and on, the animals on the basis of truth discourses about what constitutes dog wellbeing.

To explain and to elaborate on processes of agential subjectification in this context, I draw on Arun Agrawal's (2005, 17) observation that 'the question of subject-formation . . . is crucially connected to participation and practice'. Referring to the involvement of villagers in forest conservation regimes in Kumaon, India, Agrawal argues that participation in community forest protection activities leads to the internalisation of environmental norms and values and the creation of environmental subjectivities. Applying Agrawal's observation to the win-win governmentalities that pervade dog welfare efforts, I point to the immersion of AWPs in the daily challenges of implementing dog welfare programs that remain within the boundaries of the welfare episteme. This immersion in mundane practices of government for epistemic conformity leads to the unquestioning acceptance of the human-orientated normative objectives of the welfare episteme in the AWPs and influences understandings of what it means to care for street dogs.

To be more specific, the entrenchment of the human-orientated norms of the welfare episteme goes alongside the deployment and embedding of discourses and practices of animal population wellbeing. As discussed earlier, the ABC/ARV program is seen as creating stable and healthy dog populations and therefore promoting the flourishing of existing and future street dogs as a collective. The conceptualisation of animal wellbeing at the level of the population allows for individual dogs to be neutered or even killed (in the case of dogs that appear aggressive or diseased) without it being considered 'harm' per se. This allows for the simultaneous pursuit of the human- and animal-orientated objectives of the welfare episteme. Agential subjectification thus can be understood as a process in which immersion in governmentalities of care that are bounded by the welfare episteme is associated with the deployment of specific practices and discourses of more-than-human wellbeing that target dog populations and are in synergy with the human-orientated norms.

## Agential subjectification and centrifugal animal welfare discourses

Agential subjectification can be seen sharply in the transition of ABC/ARV from being viewed as a strategic, make-do intervention to being promoted as a best practice in animal welfare (as opposed to animal control), and the resultant diffusion of this biopolitical mechanism beyond the domains within which it initially emerged. I argue that it is through agential subjectification that interventions and discourses of entangled harm and care become so deeply embedded in the governmentalities of dog welfare that their harmful, strategic or make-do qualities lose significance and visibility. In fact, they start being promoted by AWPs as inherently benign and beneficial in themselves.

Foucault notes that biopolitical mechanisms often serve strategic functions, and respond to specific socio-political needs (Foucault 1980, 195). But he also explains that even if such strategic compromises arise to meet particular demands, they usually end up 'being accepted at a certain moment as a principal component ... an altogether natural, self-evident and indispensable part' of the social formations they operate in (Foucault 1991c, 75). With respect to neutering, something similiar can be observed. Interviews with local animal welfare practitioners reveal that the ABC/ARV program was originally promoted as an alternative to killing as a means of dog control (see also Nolan 2006; Jackman & Rowan 2007; Krishna 2010). However, over time, neutering has come to be increasingly promoted and understood by AWPs as intrinsically beneficial to individual dogs; as one animal welfare activist explained, 'they [the dogs] are healthier, more at peace as they won't have the urge to chase and mate'.[12] Such constructions of neutering as an animal welfare practice are seen across the world, and are directed at not only stray dogs, but also stray cats, pets of various species, and zoo animals (Jackman & Rowan 2007; Palmer et al. 2012; Srinivasan 2013).

On the whole neutering is constructed as conferring two kinds of benefits. It protects individual animals from medical issues such as ovarian and testicular cancers and mating and reproduction-related problems. By preventing the birth of dogs, it protects unborn animals from the travails of life on the streets such as 'high mortality, malnutrition, starvation, disease, and abuse' (Jackman & Rowan 2007, 55). As an AWP puts it, 'we wouldn't even like dogs on the road to begin with'.[13]

This view – that street dogs lead inadequate lives and that 'dogs . . . do better under human care', as one animal welfare practitioner said[14] – reflects animal welfare concepts and ownership models of human–dog relationships that presume that the lives of dogs that are not under human care are less privileged and are marked by serious deprivation. The implication that dogs cannot exist and thrive in the absence of human ownership is belied by the historical and contemporary thriving of free-living dogs in India and in many other parts of the world. Furthermore, as Kathy Nolan (2006) points out, 'while it is true that the life-span of a street dog is less than that of well cared [sic] pet dogs, longevity is only one measure of welfare. The quality of life of the street dog that retains its autonomy may be comparable or even better than that of many pet dogs.' This is something that I would agree with after much close observation of the lives of both pet and street dogs in India and the UK.

---

12  Interview, AWP, 2010.
13  Interview, AWP, 2010.
14  Interview, AWP, 2010.

And yet, the harmful interventions of castration and ovariohys-terectomies are increasingly re-inscribed as best practices beneficial to individual dogs and as integral components of dog welfare dispositifs. These ideational transitions from neutering as a dog control strategy to neutering as an intervention that enhances animal welfare, I argue, can be attributed to agential subjectification.

Once this view (of a 'good' dog life) takes hold, the presence of street dogs is challenged not only on grounds of human health and safety, but on grounds of dog wellbeing. Countries like the UK and USA have already seen such a shift with the consequence that dogs in these countries can exist only under human care – as pets or resources to be used in laboratories or on farms. While dogs in India can currently exist in the absence of human ownership, processes of agential subjectification have the potential to materially affect the possibilities for life available to dogs. Agential subjectification can result in situations in which the elimination of street dogs is advocated for reasons of human health and safety, and also for dog wellbeing. This is an excellent example of how mechanisms of biopolitical power are 'centrifugal', and can spill over and produce effects in new domains (Foucault 2009, 45).

## Conclusion

The recent decades, arguably from the 1970s onwards, have seen much by way of public, political and academic reflection on increasingly adverse human impacts on non-human animals and nature. Equally, there has been significant reflection on human obligations to non-human Others, and concrete actions that attempt to materially address these. The Anthropocene is thus marked by the intersection of openly articulated concern for animals and the intensification of human ways of life that are harmful to non-human wellbeing. The divergent normative objectives of the welfare episteme seen in the case of Bengaluru is one instance of this intersection; the analyses in this chapter have argued that it is this intersection that is productive of biopower in the street dog welfare dispositif and the associated intertwining of harm and care.

In making these observations about biopolitical power in spaces of more-than-human social change such as animal (street dog) welfare, this chapter draws attention to processes of agential subjectification in those humans who advocate for, and act on behalf of, animal Others. In particular, it points to some of the possible consequences of extended and extensive involvement in juggling various socio-political imperatives and highlights how ideas of what constitutes animal wellbeing can come to be influenced by factors that are not entirely to do with the wellbeing of animals. Agential subjectification is not universal, as can be seen in the doubts expressed by at least some AWPs about the benefits and impacts of neutering for the animals. It is equally clear that such biopolitical mechanisms have gained currency and continue to be promoted and diffused as best practices – and not merely strategic compromises that are meant to be stopgap solutions for a less-than-perfect situation.

The analyses developed here invite debate and reflection from those involved in creating equitable more-than-human presents and futures. A question they raise is whether spaces of social change, whether intra-human or more-than-human, are always already marked by the play of non-benign biopower because of the coming together of continuity and change. The Anthropocene is marked by the privileging of human interests as well as criticism of the more-than-human impacts of human exceptionalism. Does this simultaneity result in the re-articulation of human interests in the pursuit of animal wellbeing? In posing this question, this chapter provokes critical consideration of the complicated contours of more-than-human social change in the Anthropocene; this is especially with regard to the subtle and diverse ways in which human exceptionalism is re-articulated in domains of improvement and reform such as animal welfare and environmentalism.

## Works cited

Agrawal A (2005). *Environmentality: technologies of government and the making of subjects*. Durham, NC and London: Duke University Press.

ASPCA (2012). How will neutering change my dog? American Society for the Prevention of Cruelty to Animals. Retrieved on 29 April 2015 from http://www.aspca.org/Pet-care/virtual-pet-behaviorist/dog-articles/how-will-neutering-change-my-dog; See also www.aspca.org/Pet-care/virtual-pet-behaviorist/dog-articles/how-will-spaying-change-my-dog.

Beary H (2007). Stray dogs killed in Bangalore. BBC News. Retrieved on 29 April 2015 from http://news.bbc.co.uk/1/hi/world/south_asia/6446517.stm.

Chief Health Officer WP 427 (2007). Affidavit of the Chief Health Officer, BBMP, Writ Petition 427/2007, B Krishna Bhat and Union of India & Others, High Court of Karnataka at Bangalore.

Cole M (2011). From 'animal machines' to 'happy meat'? Foucault's ideas of disciplinary and pastoral power applied to 'animal-centred' welfare discourse. *Animals* 1(1): 83–101.

Compassion Unlimited Plus Action WP 427 (2007) Response of CUPA, Writ Petition 427/2007, B Krishna Bhat and Union of India & Others, High Court of Karnataka at Bangalore.

Crutzen P & Stoermer EF (2000). The 'Anthropocene'. *Global Change Newsletter* 41:17–18. Retrieved on 29 April 2015 from http://www.igbp.net/download/18.316f18321323470177580001401/NL41.pdf.

Dean M (2010). *Governmentality: power and rule in modern society*. 2nd edn. London: Sage.

*Deccan Herald* (2007a). BBMP cries halt to dog hunt, for now. *Deccan Herald*, 10 March.

*Deccan Herald* (2007b). Eye for an eye policy won't work. *Deccan Herald*, 8 March.

Duijzings G (2011). Dictators, dogs and survival in a post-totalitarian city. In M Gandy (Ed). *Urban constellations* (pp145–48). Berlin: Jovis.

Elliot M (2008). Neutering: the pros & cons. *Dogs Monthly Magazine*, November. Retrieved on 4 May 2015 from http://www.prescotpages.co.uk/documents/MarkElliott_NeuteringProAndCon.pdf.

Foucault M (1977). *Discipline and punish: the birth of the prison*. London: Penguin Books.

Foucault M (1980). The confession of the flesh. In C Gordon (Ed). *Power/knowledge: selected interviews and other writings 1972–77: Michel Foucault* (pp194–228). New York: Pantheon Books

Foucault M (1991a). Governmentality. In G Burchell, C Gordon & P Miller (Eds). *The Foucault effect: studies in governmentality* (pp87–104). Chicago: The University of Chicago Press.

Foucault M (1991b). Politics and the study of discourse. In G Burchell, C Gordon & P Miller (Eds). *The Foucault effect: studies in governmentality* (pp53–72). Chicago, IL: University of Chicago Press.

Foucault M (1991c). Questions of method. In G Burchell, C Gordon & P Miller (Eds). *The Foucault effect: studies in governmentality* (pp73–86). Chicago, IL: University of Chicago Press.

Foucault M (2002). *The archaeology of knowledge*. London & New York: Routledge.

Foucault M (2003). *Society must be defended: lectures at the College de France 1975–1976*. New York: Picador.

Foucault M (2008). *The history of sexuality. Vol. 1: The will to knowledge.* Camberwell, Vic.: Penguin Books.

Foucault M (2009). *Security, territory, population: lectures at the College de France 1977–78*. Basingstoke, UK & New York: Palgrave Macmillan.

Francione GL & Garner R (2010). *The animal rights debate: abolition or regulation?* New York: Columbia University Press.

Franklin A (1999). *Animals and modern cultures: a sociology of human–animal relations in modernity*. London: Sage.

Garner R (2004). *Animals, politics and morality*, 2nd edn. Manchester, UK: Manchester University Press.

Hobson K (2007). Political animals? On animals as subjects in an enlarged political geography. *Political Geography* 26(3): 250–67.

Holloway L & Morris C (2012). Contesting genetic knowledge-practices in livestock breeding: biopower, biosocial collectivities, and heterogeneous resistances. *Environment and Planning D: Society and Space* 30(1): 60–77.

Igoe J, Neves K & Brockington D (2010). A spectacular eco-tour around the historic bloc: theorising the convergence of biodiversity conservation and capitalist expansio. *Antipode* 42(3): 486–512.

Jackman J & Rowan AN (2007). Free-roaming dogs in developing countries: the benefits of capture, neuter and return programs. In DJ Salem & AN Rowan (Eds). *The state of the animals IV: 2007* (pp55–78). Washington, DC: Humane Society Press.

Johnson E, Morehouse H, Dalby S, Lehman J, Nelson S, Rowan R, Wakefield S, & Yusoff K (2014). After the Anthropocene: politics and geographic inquiry for a new epoch. *Progress in Human Geography* OnlineFirst. doi:10.1177/0309132513517065. Retrieved on 19 May 2015 from http://phg.sagepub.com/content/38/3/439.

Kotchen MJ & Young OR (2007). Meeting the challenges of the Anthropocene: towards a science of coupled human-biophysical systems. *Global Environmental Change* 17(2): 149–51.

Krishna CS (2010). Dog population and rabies control in India. Paper presented to Responsible Dog Ownership in Europe: Conference on Canine Overpopulation in Europe and Sustainable Solutions Strategies, 4–5 October 2010, Brussels. Retrieved on 4 May 2015 from http://tiny.cc/53jeyx.

Legg S (2005). Foucault's population geographies: classifications, biopolitics and governmental spaces. *Population, Space and Place* 11(3): 137–56.

Liptak JM, Dernell WS, Ehrhart N & Withrow SJ (2004). Canine appendicular osteosarcoma: diagnosis and palliative treatment. Retrieved on 29 April 2015

from http://animalcancersurgeon.com/Review_Articles_files/
Compendium%202004%20OSA%20I%20diagnosis%20and%20palliation.PDF.

McHugh S (2004). *Dog*. London: Reaktion Books.

MoEF WP 427 (2007). Response of the Ministry of Environment and Forests,
Union of India, Writ Petition 427/2007, B Krishna Bhat and Union of India &
others, High Court of Karnataka at Bangalore.

Mol PJ, Sonnenfeld DA & Spaargaren G (Eds) (2010). *The ecological modernisation
reader: environmental reform in theory and practice*. London & New York:
Routledge.

Munro L (2012). The animal rights movement in theory and practice: a review of
the sociological literature. *Sociology Compass* 6(2): 166–81.

Murdoch J (2003). Geography's circle of concern. *Geoforum* 34(3): 287–89.

Nolan K (2006). 'Street dog' population control. Vetwork UK. Retrieved on 29
April 2015 from http://www.vetwork.org.uk/abc.htm.

Palmer C, Corr S & Sandoe P (2012). Should we routinely neuter companion
animals? *Anthrozoos* 25(Supplement): S153–72.

Philo C & Wilbert C (2000). Animal spaces, beastly places: an introduction. In C
Philo & C Wilbert (Eds). *Animal spaces, beastly places: new geographies of
human–animal relations* (pp1–34). London and New York: Routledge

Rabinow P & Rose N (2006). Biopower today. *BioSocieties* 1(1): 195–217.

Reichler IM, Hubler M, Jochle W, Trigg TE, Piche CA & Arnold S (2003). The
effect of GnRH analogs on urinary incontinence after ablation of the ovaries
in dogs. *Theriogenology* 60(7): 1207–16.

Schmidt K (2011). Concepts of animal welfare in relation to positions in animal
ethics. *Acta Biotheoretica* 59(2): 153–71.

Srinivasan K (2013). The biopolitics of animal being and welfare: dog control and
care in the UK and India. *Transactions of the Institute of British Geographers*
38(1): 106–19.

Srinivasan K (2014). Caring for the collective: biopower and agential
subjectification in wildlife conservation. *Environment and Planning D: Society
and Space* 32(3): 501–17.

*Times of India* (2007). Living with animals reflects tolerance. *Times of India*, 25
March. Retrieved on 29 April 2015 from http://timesofindia.indiatimes.com/
edit-page/Living-with-animals-reflects-tolerance/articleshow/1800191.cms.

Twine R (2010). *Animals as biotechnology: ethics, sustainability and critical animal
studies*. London: Earthscan.

WP 4920 (2007). Writ Petition 4920/2007, George Mathew & Others and Union of
India & Others, High Court of Karnataka at Bangalore.

Youatt R (2008). Counting species: biopower and the global biodiversity census.
*Environmental Values* 17(3): 393–417.

# 12

# Wild elephants as actors in the Anthropocene

*Michael Hathaway*

Here I explore the issue of non-human agency in the age of the Anthropocene.[1] The notion that non-human animals have agency is just one of an increasing number of challenges to a long-enduring Western conceptual framework that views non-human animals and humans as

Hathaway M (2015). Wild elephants as actors in the Anthropocene. In Human Animal Research Network Editorial Collective (Eds). *Animals in the Anthropocene: critical perspectives on non-human futures*. Sydney: Sydney University Press.

1 The term 'Anthropocene' has a recent history, being coined in the year 2000 by Paul Crutzen and Eugene Stoermer. It suggests that humans have become the most significant ecological force on the planet, akin to a geological time period, such as the advance of glaciers in one of the numerous Ice Ages that have carved the Earth. The concept that humans have deeply shaped the global environment goes back further than 150 years, when Karl Marx argued that 'nature, the nature that preceded human history . . . no longer exists anywhere (except perhaps on a few Australian coral-islands of recent origin)' ([1845]1970, 63). Almost 30 years later, in 1873, an Italian geologist, coined the term the 'anthropozoic era', but this term did not gain much traction. The current term 'Anthropocene' is primarily used by geologists and scientists interested in climate change, but has caught on with a wider public, including scholars across a range of disciplines, such as those included in this book. In part, it likely gains an audience because it reinforces an existing sense that we live in an era of environmental crisis that is precipitated by human actions.

intrinsically different. In this anthropocentric frame, non-humans are regarded as 'passive' counterparts to 'active' human subjects, the latter with conscious intent and agency and the former without. As a result, non-human species are posited as the victims of human actions: domestic animals are turned into living commodities, and wild species suffer from human destruction of their habitat. A number of animal studies scholars consider the concept of non-human agency in one of two frames, by either lamenting the condition of animals as victims, or celebrating their acts of 'resistance' against the human world. I was given cause to examine the concept of non-human agency in these terms during long-term anthropological fieldwork in Southwest China. My family and I spent a year in a rural village where villagers live alongside the country's last remaining herds of wild elephants. During that time my conversations with the villagers and observations of the human–elephant entanglements in that area led me to a view of animal agency that is agnostic about the question of conscious intent, but curious about what might be called 'distributed agency' or 'cumulative agency', which explores how accumulation of actions over time and by many individuals has effects in the world. In this chapter I pursue a sense of agency that is not predicated on unusual actions by specific individuals, but that recognises elephants as place-makers and historical actors whose behaviours are in dynamic flux.

I examine the concept of the Anthropocene, and then explore how some important ideas about non-human agency might address some of its potential shortcomings for the effects that anthropogenic factors have on non-human species. I then discuss two examples of human–elephant entanglements in China, one in which the relationships between villagers and elephants are mediated by a powerful conservation organisation, and one where elephants have impacted the design and the operation of a major highway. These examples illustrate how we might understand non-human agency differently. I consider these relationships in the context of the Anthropocene for several reasons. First, I am concerned that common understandings of the term 'Anthropocene' might be problematic by reinforcing forms of anthropocentrism that limit our reimagining of the place of humans among life on Earth (Dibley 2012). Second, I use recent examples that illustrate how the lives of elephants are historical and not merely evolutionary: their lives are part of a range of predicaments that they shaped and, in turn, shape their contemporary lives.

The editors of this book ask how the concept of the Anthropocene might 'enable, provoke and frame discussions of multispecies responsibility and justice'. Typically the notion of the Anthropocene is anthropocentric, viewing humans as the main, or sole actors, in shaping the Earth. I ask: What do we lose by missing an understanding of our connections with other species and non-human forces in the world? My perspective is influenced by the rise of scholarship on the 'more-than-human' (sometimes also called 'posthumanism') and in particular, Jane Bennett's work on 'vital materialism', which explores the agency of non-humans, both living and non-living (2007). Bennett asks how different understandings of agency might challenge human understandings of living in the world. Even mundane activities, such as eating, might be understood in a different light through Bennett's concept of vital materialism (2007). For example, Bennett argues that our 'conquest model of consumption':

> disregards the effectivity of not only animal bodies, but also the 'bodies' of vegetables, minerals, and pharmaceutical, bacterial or viral agents. It presents nonhuman matter as merely the environment for or the means to human action. But does there not exist, as the notion of a viral agent suggests, a form of agentic capacity not restricted to the human actor, a potentiality within materiality per se? (2007, 133)

Bennett's aim is to reveal a world more lively than we might imagine, challenging commonplace notions of humans as the sole actors on the planet, as singular dominators (2007). Bennett's work raises cautions that the mainstream understanding of the present era as the Anthropocene may unintentionally reinvigorate a notion of human dominance. The environmental crises of the Anthropocene require us to take responsibility for our impacts upon the non-human world, yet the mainstream perspective may also reinforce a sense of humans as masters and caretakers of all, not co-participants in multispecies worlds. But what if we were to extend the concept of agency beyond the human and recognise non-human species in similar terms? How would this affect our understanding of human–animal relations in the Anthropocene?

## Species interconnections

A legacy of scholarship on human–animal relations has often focused on animals as either victims or resistors. Such scholarship is now expanding in important ways, imagining non-humans not just in these terms but as intimately interconnected with human lives. Geographers Sarah Whatmore and Lorraine Thorne argue that we might better understand these entanglements by thinking more about forms of agency than cross-species boundaries. Borrowing from actor–network theory (ANT), as conceived by Bruno Latour, they use the term 'actant' to describe 'something that acts or to which activity is granted by others' (2004, 75). Such a definition applies to and encompasses all living beings and expands upon the more typical connotation of agency, which requires conscious intentionality and motivation, qualities often presumed to be possessed only by humans. Whatmore and Thorne present a networked conception of human–animal relations as:

> a relational achievement spun between people and animals, plants and soils, documents and devices, in heterogeneous social networks that are performed in and through multiple places and fluid ecologies. (1998, 437)

Viewing such networks as fluid rather than static, and species as interconnected rather than separate, emphasises the negotiations that take place in such relations.

Whatmore and Thorne's work is part of a broader shift in views of human–animal relationships based on ANT models. Ironically, although many ANT-influenced scholars tend to imply that all actants in a network are equally agential, and aim to displace humans as the sole actors, actual studies of networks still tend to foreground humans as the main agents in creating and in expanding networks. Non-human agents can become part of these networks as a result of their involvement with humans, but their activities outside of these networks and their status of agents are ignored. Thus, while humans are not the sole actants, they are always in the centre in these accounts, with non-humans unwitting participants in these networks.

One of the few instances that show the 'failures' of such readings of non-human agency in ANT literature occurs in work by Michel Callon, who described the 'refusal' of scallops to be cultivated in St Brieuc Bay in France; beings that did not grow where humans intended (1986). Yet, such examples of refusal are rare in ANT accounts, and while non-humans are described as 'actants' in the same language as humans, their active presence is notably diminished by comparison. Tim Ingold's approach might ameliorate such a tendency in ANT. Ingold argues that while most accounts of animals portray them as moving across an inert landscape and seeking existing habitat niches, many species actually create these niches through actions such as nest-building, damming creeks and hunting, each of which brings changed landscapes into being through an interactive and iterative process (2011, 78). Such a perspective fosters two kinds of enlivening: it both animates a landscape typically seen as fixed, and views animals as actively shaping their worlds, and not as living a passive existence.

We understand human and non-human relations, and in particular interactions with other animals, through two narrative frameworks that have dominated accounts in animal studies: those of victimisation and of resistance. Scholars such as Clare Palmer (2006), Sarah E McFarland and Ryan Hediger (2009) remind us that much animal studies scholarship portrays animals almost solely as victims. Jason Hribal agrees, and suggests further that even in many animal studies' most influential theories:[2]

the animals are not seen as agents. They are not active, as laborers, prisoners, or resistors. Rather, the animals are presented as static characters that have, over time, been used, displayed, and abused

---

2   Hribal includes work such as Harriet Ritvo's *Animal estate: the English and other creatures in Victorian England* (1987), Nigel Rothfels's *Savages and beasts: the birth of the modern zoo* (2002), Louise Robbins's *Elephant slaves and pampered parrots: exotic animals in 18th century Paris* (2002), and Virginia Anderson's *Creatures of empire: how domestic animals transformed early America* (2006).

by humans. They emerge as objects – empty of any real substance.
(2007, 102)

Hribal's assertions are intriguing, and I would argue that the focus on victimisation that is common in many accounts of subaltern human history leads to an analysis where the main subjects are relatively passive objects without the scope to be active agents. Yet, Hribal's alternative focus on 'resistance' is also problematic for several reasons. Hribal mainly studies animals used for labour, and has discussed cases where elephants and lions working in the entertainment industry have struck back at their trainers. He argues that these behaviours constitute resistance. This concept also arises in the work of Philo and Wilbert, who also conflate animal agency with resistance (2000, 14). In asking how we might notice such acts of resistance, for example, Philo and Wilbert suggest that one example is the act of 'trespass' where animals appear in places that humans have tried to keep them from (2000, 13). Other studies of resistance borrow from exemplary work by political scientist James Scott, including his book *Weapons of the weak: everyday forms of peasant resistance* (1985). Scott's accounts have been persuasive and stimulated much important work on writing a 'history from below' (historical accounts from subaltern perspectives) and show the agency of peoples previously assumed to lack political awareness and action. Yet, as I have written elsewhere, the academic concept of resistance greatly simplifies subaltern consciousness, reducing complexity to a singularity, assuming reaction more than action, and imposing an analytic stance that is presumed rather than explored (Hathaway 2013). In a resistance framework, many diverse actions are often conflated and seen as evidence of resisting, whether this is for humans and non-humans. I do not wish to completely abandon the notion of resistance, but find other perspectives in which resistance is just one part of a much broader repertoire to explain non-human actions.

For China's herds of wild elephants, particular actions that might be regarded as 'resistance' make up a relatively small part of the animals' repertoire of actions and behaviours. As I explain, elephants trespass in places where humans do not want them, invading farmland and eating crops, crossing roads and threatening the safety of humans. Are these acts of resistance? They certainly show agency, but if we consider agency only as forms of resistance, we are brought back to anthro-pocentric notions where resistance is only seen in relationship to human practices and actions, and not to those of other species and other networks and entanglements. The social worlds of many wild animals are much more strongly mediated, in a daily way, by encounters with other animals, of their own and other species, both animals and plants. Does it make sense to talk of an elephant's resistance to species other than humans, or does resistance necessitate working against human intentions? Would such a definition re-emplace humans in the centre of the analysis? Elephants live in landscapes shaped by humans, but they also play an ongoing role in configuring these landscapes. They do so as voracious omnivores, shaping the forests around them as they move through them. They also configure landscapes as political subjects. Humans who live in proximity with elephants, such as the villagers I spent time with in Southeast China, accommodate the animals' presence in many ways, mediating the relationship between themselves and these large and dangerous animals. I now consider some of these mediations and how they affect the lives of both elephants and humans in the village in which I conducted my anthropological study.

## Relationships between humans and wild elephants in Yunnan Province

For millennia, humans have interacted with China's wild elephants in numerous ways. In the past the Chinese controlled and manipulated elephants, capturing and training them for use in royal processions, or as formidable mounts in armies (Elvin 2004). In both the past and the present day they have been killed for their ivory. In these scenarios, we mainly view humans as the actors and elephants as the victims. Yet in other contexts, humans appear as the victims of elephants' actions. In modern day China, elephants often seek out human communities as they roam the tropical rain forests that skirt the agricultural fields of Yunnan Province, near the Mekong River. While neighbouring nation-states such as Myanmar, Laos and Thailand each have several thousand elephants, both wild and captive animals, in China elephants are considered endangered, with only about 200 to 300. A few millennia ago, when the climate was warmer, China's elephants lived more than 1000 km to the north, as far as Beijing, but gradually retreated to the south as the climate changed. Now, they occupy only a small area, less than one-third of the size of Tasmania (Elvin 2004).

My understandings of elephants changed dramatically when I spent a year between 2000 and 2001 living with my family in a village called Xiao Long, located within the Xishuangbanna Nature Reserve. Whereas I once believed that elephants would naturally steer clear of humans, I learnt that China's wild elephants live in close proximity to human communities. The villagers with whom I lived with taught me that elephants loved to eat their crops, and actively sought out agricultural fields to raid. My neighbours told me many stories about elephants destroying an entire year's harvest of grain upon which they depended. One night's work by elephants could mean that a family faced hardship until next year's grain harvest. The family might try to make do with less food, eating smaller meals until the next year's crop, or they might be forced to buy food, creating a debt that would take much longer than a year to pay off.

I also learnt that dogs were important mediators of the relationships between humans and elephants. Most of the village households had one or two dogs. Many were quite sensitive and detected nearby elephants, alerting the villagers to their presence. Elephants can be surprisingly difficult to see or to hear when they are in thick forest, so having a dog with those skills is good insurance against being surprised by a rampaging animal. Sometimes, dogs could successfully scare an elephant, forcing it to return to the forest, thereby saving their owner's crops. However, sometimes the dogs' incessant barking could anger an elephant. Houses in the village are made of bamboo with a grass thatch roof, and through time a number of houses had been destroyed by an angry elephant chasing after a barking dog that had retreated inside to escape being attacked. The villagers told me that if a dog chased by an elephant ran into my house that I should run out – with the dog if I could – to save the house and myself.

Evidently the different behaviours of the elephants greatly impact upon the villagers' livelihood and safety, and one of the ways the villagers have tackled the problem was by shooting warning shots into the air to scare the elephants. For the villagers, this was necessary at times when the elephants presented the greatest threat to their livelihood or their safety. However, the Xishuangbanna Nature Reserve Bureau was concerned by this behaviour and orchestrated a massive gun confiscation campaign with the help of local police, including one while I was living there in the year 2000. Almost all of these guns were muzzle-loaders, and few were really capable of killing an elephant, for they possess thick skin and bones and are strikingly resistant to musket balls. The few convicted poachers who were capable of killing elephants had high-powered military guns, but villagers never admitted to killing elephants with their own muzzle-loaders. I later heard stories from biologists about coming upon an old elephant skeleton with dozens of lead balls among the bones, testimony to their long-term antagonisms with farmers and to their own power of survival. In addition to confiscating guns, the bureau staff worked in alliance with government officials, who tried to reduce the sales of lead blocks, commonly used for making musket balls. Most local men owned a gun, using it for hunting or for simply defending their crops, especially when the rice and corn ripened, a time of great concern when not only elephants, but monkeys, wild pigs, forest deer and other creatures ventured into the fields to eat their fill. People told me that soon after their guns were taken away, elephants gained a newfound confidence in entering villages to raid crops and even attacked groups of people walking, driving their tractors, or riding motorcycles on forest trails. They described this as the 'revenge of elephants'. Many people became especially cautious about walking at dawn and at dusk, reducing the amount of time that they worked in the fields as they now went later and returned earlier. Thus their lives were affected both by the elephants that shared their territory, and by the involvement of police and officials from the Xishuangbanna Nature Reserve Bureau. Deeply mediated by their existence in a network, these humans and elephants were entangled in ongoing dialectical relationships. The tenor of these engagements, in this village and thousands of others affected by the elephants, creates the social context from which Chinese officials, wildlife experts and others approach what is often termed the 'elephant issue'. I now explore this idea in greater detail with reference to the effects that a global conservation organisation and the development of road infrastructure have had on the region.

## Involvement of the World Wildlife Fund

In the late 1980s the World Wildlife Fund (WWF) – the world's largest transnational conservation organisation – carried out its first operations in Yunnan Province.[3] It originally set up operations in China focused on the conservation of the panda – the organisation's mascot – and they planned to be the first conservation non-government organisation to work in China after it 'opened up' to the West. It soon initiated a number of projects that were to do with the elephants, creating China's first ecotourist site – Wild Elephant Valley – that allowed tourists to watch elephants in their own habitat. WWF also tried to protect some villages from crop-raiding elephants by building electric fences. The latter effort was part of a deal that they made with villagers: they would keep elephants away from the fields, and villagers would cease to carry out slash-and-burn agriculture in the surrounding forest, destroying the animals' habitat.[4]

During the first stages of the WWF projects the human inhabitants of Yunnan Province believed that this well-financed effort – initiated by a powerful Western organisation with access to resources and sci-

---

3   It is interesting to note that WWF, based in Geneva and mainly financed by Europeans, only came to China for the first time in the 1980s as likely the first Western nature conservation effort in the country. Thus, the Europeans' conservation efforts in China were nearly one century after working to create and to enforce conservation policy in Africa and other regions around the world. The African conservation efforts were often aimed at excluding local peoples from hunting, and allowing colonials to hunt, rather than conservation across the board. WWF began in 1962 with its panda mascot, but despite interest to work in China, was not invited to visit by Chinese leaders until a bamboo flowering crisis that killed the main food of pandas, triggered a wide-scale panda famine.
4   In the 1980s, slash-and-burn was considered harmful to the tropical rainforest as it was no longer seen as gloomy jungle but a place of great ecological value and a threatened ecosystem, described then as the 'lungs of the earth'. Many accounts of the tropical rainforest saw slash-and-burn as the great threat to rainforest destruction, and WWF was only one of many organisations that aimed to eliminate its practice among the hundreds of millions of farmers in the tropics. Only later did scientists challenge their interpretation on slash-and-burn, seeing it, when practised in small patches by local farmers, as providing greater diversity, an ecological mosaic than enhanced biodiversity, including improving the habitat of wild elephants.

entific knowledge – would solve the longstanding problems associated with human–elephant conflict. However, within a few weeks, elephants learnt how to destroy the electric fencing by wielding heavy branches in their trunks that they used to break the wires. Despite the WWF and the villagers' repeated efforts to fix the fencing, they could not thwart this behaviour. When villagers told me this story, years after the fact, I told them that similar electric fencing seemed to work in Africa. Some of the men, who loved telling elephant stories of danger and destruction, chuckled, 'Well our elephants are Chinese, and they are smarter than African elephants.' Further, there was a sense of pride that their elephants could outwit Western technology and expertise. Those responses showed me the complex nature of the interactions between humans and elephants in that region. Villagers had also told me stories about elephants being clever and highly talented, able to do things requiring dexterity such as using their trunks to shuck the husk from ear of corn. They also described the elephants as 'naughty', rolling in fields of ripe corn or rice, seemingly for the destructive pleasure of it. In hearing these accounts I realised that as well as a sense of pride for the animals' intelligence, there was a fondness for elephants that existed and a great respect for the animals' power and ongoing fears of the impact that they could have on the fields and village. This was a complex relationship, mediated by the efforts of international conservation organisations and the Chinese state, but that also predated engagements with such institutions. We might suspect that rural villagers would be relatively helpless in the face of such elephant incursions, but would likely expect that major construction projects carried out by the Chinese state would not only probably ignore the presence of elephants in their design, but would be unaffected by elephant actions. In the case of the Sixiao Highway, however, both assumptions would prove untrue.

## The Sixiao Highway

The presence of wild elephants has affected the design and use of a new road, called the Sixiao Highway, completed in 2006. The road, partly financed by the Asian Development Bank, was part of a plan to link Yunnan's capital of Kunming with Bangkok, Thailand. The new highway has affected the elephants' seasonal circuits, as they search out human-grown crops, patches of tender bamboo, mineral salts, dust and gentle river pools for bathing. Of all of the hundreds of animal species living in these tropical rainforests, elephants were of most concern to the highway planners. In conversation with biologists the planners learnt that the new highway would block many of the elephants' long-standing migratory paths as they often travel in vast circuits, in part stimulated by the seasonal shifts between the rainy and dry seasons, and also in search of mineral-rich salt licks. In crossing or walking across the highway they could have potentially been physically hurt, either as the road was being built, or as a result of traffic accidents that also threatened the safety of drivers (Pan et al. 2009). In light of this, planners changed the original course of the road's path, and built 16 tunnels for elephants and other animals underneath the highway, mainly at existing elephant pathways. To further dissuade elephants and other animals from crossing the road, they also built a stout, two-metre high metal fence along certain areas (Pan et al. 2009).

The old highway had been built on the cheap in the early 1950s. Mainly created with manual labour, the old highway had few bridges and no tunnels. It was rarely even two lanes wide and drivers often moved slowly and carefully, averaging below 30 km per hour. The highway was never fenced and followed winding river valleys. Elephants crossed it freely, and sometimes even walked along it, especially when the traffic was light. Elephants might have changed their own paths during its construction and afterwards, but we will not know, as there was no research on elephant routes carried out at the time. In fact, when the old highway was planned, many officials in Beijing thought that China's elephants were extinct, but in the early 1950s a team of researchers found remnant herds, a cause of great excitement in the scientific community (Hathaway 2013).

The old highway mostly followed a centuries-old caravan route. The road was not designed for vehicles with wheels, but for caravans of human porters and pack-laden mules. It had many hewn stone steps, some of which were chiselled into the rock of steep slopes, and elephants typically avoided these areas, especially when the tread was too narrow or too steep for their large feet and ambling gait. Although traffic along the caravan highway was busy at times, with a vigorous trade in tea, silks, opium and other luxury goods, these caravans moved slowly, giving elephants plenty of time to move away from people if they so chose. Elephants do not instinctively fear humans, but learn to distinguish situations in which humans may or may not threaten them (Bates et al. 2007).[5] As well as some of the terrain being inaccessible to the elephants, the caravan route and the old highway would have affected the elephants' movements, depending on their avoidance behaviours about humans.

The new highway has further changed the elephants' routes through the region. They have abandoned some of the old paths by not walking through tunnels built for them, creating new routes by walking across bridges designed for human traffic, and have found new ways to cross the highway where planners have not installed fences. In the past the elephants rarely showed much fear of slow-moving cars and trucks on the old road. Today drivers on the new highway are curious about elephants and may drive slowly or may stop if and when they see elephants. However, many drivers travel very fast and this causes numerous problems for both elephants and humans. There have been serious problems on the new highway due to drivers blaring their horns at elephants on the road. This has caused some of the animals to become anxious and to damage cars, threatening the safety of passengers. Generally, local police respond quickly to reports of elephants on the road, and attempt to drive them off quickly. As traffic on the highway has increased, elephants appear to have become more nervous around vehicles and have been shifting from diurnal to nocturnal their patterns of movement in order to cross the road during lighter traffic at night.

These examples show that the presence of elephants in China shapes a wide range of human engagements. As conservation efforts grow, the elephants have responded by 'becoming' bolder, seemingly more likely, for example, to raid crops in villages where guns have been confiscated, and efforts to protect crops with electric fences have failed spectacularly. But I am not content to categorise these activities as resis-

---

5 For example, as Bates explains, in Kenya elephants seem to distinguish between two ethnic groups, one that threatens them and one that does not, differentiating the groups by smell and by sight (Bates et al. 2007).

tance and the only evidence of their agency. Instead, we might cast our eyes more broadly, to look at all forms of elephant behaviour as effectual, rather than allocate acts into the categories of resistance and non-resistance. In some ways elephants do as expected (such as using the tunnels specifically designed for them to walk safely under the road), but also deviate from expectations (such as using bridges designed for human pedestrians). There are also direct examples of elephants changing their behaviour as a result of human activities, such as the shift from diurnal to nocturnal movements. In order to account for this array of agential behaviour, we need to have a notion of agency that is iterative, historically dynamic, and explores the kinds of changes generated through interactions between groups of humans and non-humans. As a heuristic exercise to consider these dynamics, I now introduce the notion of alternate histories and animal governmentalities.

## Alternate histories and animal governmentalities

As the elephants of Yunnan Province are neither tagged nor intensively studied, we do not know enough about the lives of individual elephants in China to construct detailed biographies. This has led me away from trying to understand agency as acts by individuals performing notable feats or actions that have garnered human attention, towards the understanding of agency as a more diffuse property, generated by groups. This is akin to what Jeremy Prestholdt refers to as 'cumulative agency' (2007) and to what Jane Bennett calls 'distributed agency' (2009), both of which are seen to locate forces in the world as being created through groups or networked actants. One might understand these concepts through the notions of 'alternate histories' and 'animal governmentalities'. As a thought experiment, we can imagine an alternate history where elephants had never existed or had been eliminated in China. How would it be different? In China, there was actually a high likelihood that elephants would have become extinct. In the last 1000 years elephant habitats in China have shrunk to smaller than two percent of their former size, barely supporting the 200 to 300 animals that remain today. Briefly compare their fate with tigers: into the 1950s, the populations of tigers and elephants were possibly roughly the same, but after the government categorised tigers as 'vermin' and offered a bounty for their skins, every tiger was killed. The same could have happened to elephants, as they threatened farmers' livelihoods and people's safety. During the Mao era, wild animals were generally seen in utilitarian terms for their monetary value, or were otherwise eliminated as pests. Thus the fact that a small population of wild animals still exists in China is a somewhat remarkable outcome. I imagine that the elephants' charisma and behaviour played a role in this outcome, and their long history of being a valued object of Chinese royalty and a prized zoo specimen. Therefore, imagining an alternate history in which China's elephants had become extinct can reveal the myriad ways in which the continuing presence of these animals shape the social and natural landscape today. Elephants not only seek out food and cross highways but they are also significant parts of vast networks that they share with people who aim to study them, advocate for them and interact with them in public and private spaces like transport routes, fields, villages and homes. The continuing presence of elephants was a critical factor in WWF going to that province, and continues to foster the strong presence of federal funding and support for conservation, all forms of what one might call a distributed form of agency where people and Others are entangled in each others' lives.

A second way to think about these networks is on the governmentalities that affect the lives of elephants. In part this is what Whatmore and Thorne are referring to the networks of documents and devices that engage wildlife (1998, 437), but it is more than that. Compared to China's forest elephants, African savannah elephants live within far more contentious and overlapping systems of governance. In Africa, elephants are surveyed with low-flying planes or are fitted with GPS trackers. China's forest elephants would not be visible from an aerial survey. To my knowledge, no one in China has ever placed a tracker on one of those animals, or has conducted a systematic land survey. In Africa, many elephants are relatively contained within nature reserves that act as habitat 'islands', patrolled by armed guards. Biologists frequently assess the 'carrying capacity' of nature reserves and enter into highly contentious debates about the economics and moral acceptability of different techniques to 'cull' elephants when the populations rise beyond the amount of available food (Thompson, 2002).

Some advocate for elephant birth control using the drug Depo-Provera, while others believe in inoculating elephants against disease. These forms of governance, aimed to increase some populations and to decrease others, monitor their food supplies, and so forth, are denser in many African countries (although these differ greatly between countries) than in China, in part because a longstanding passion for wildlife in Africa is linked to centuries of colonial rule and postcolonial relations.

In comparing African and Chinese elephant governmentalities we should not follow the frequent assumption in Foucauldian studies that the creation of statistics and plans to manage populations are always effective (Foucault 2009). Despite such active monitoring in Africa, thousands of elephants are killed by poachers. In China, however, the number of poached elephants is extremely low, despite the intensity of violent interactions between rural villagers and elephants. Since early in the 21st century, China is now held up by some elephant welfare advocates such as the UK's ElephantAid as a model country for elephant conservation. They point out that China is one of the few countries that follow a 'no kill' policy: it does not sanction the destruction of any elephant for any reason, even those known as 'rogue animals' that may be particularly violent. In India, for example, leading elephant researchers might advocate such killing, while in China no researcher appears to take such a stance, and there seems little sympathy for such a position.

China has often been considered by animal rights advocates as one of the world's worst offenders in terms of animals' treatment in zoos and farms, yet the story of its wild elephants shows a different side of state power. The Chinese state is an active participant in elephant advocacy, at least for its own elephants (its relatively lax position, at least until recently, on the massive imports of African ivory is another issue, as China becomes the world's biggest consumer of illicit ivory). While it may earn kudos from urban and international elephant lovers, at a regional scale the state's 'no kill' policy raises challenges to state legitimacy as rural people in elephant territory increasingly complain that the state is more interested in elephant welfare than in human welfare. During my fieldwork I found that people in rural areas complained that it was the government's interest in elephants that prompted the gun confiscation campaign. As previously mentioned, they also said that the elephants have changed their behaviour once they knew that the villagers were unarmed, and are now seeking revenge. I was able to compile surprising statistics from a Chinese government report – only available on the internet for a short time before being removed – that stated that elephants had killed more than 60 people in the course of a five-year period. This number is especially high when as there are probably only 200 elephants in a relatively small area. From my preliminary research China's elephants appear to be the most dangerous group of elephants in the world (Hathaway 2013). Thus, the actions of these elephants mean the stakes are high in China. Public sentiment demands government action to address the problems faced by rural people who are increasingly at the mercy of wild elephants that raid their crops and attack their villages. Elephant activity has become an intensely political issue for a country that is already worried about social stability.

Surprisingly, while elephants seem to cause many problems for the Chinese government, so far state officials and biologists are working to expand the territory of elephants and to increase their numbers, not control or limit them. Chinese government officials have travelled to Laos, working to create international nature reserves that are designed specifically to benefit elephants, and to allow them greater habitat and corridors to expand their range. Overall, there is a great deal of work going on, all due to the presence of elephant's and their importance in China.

Compared to centuries ago today, at a global level there are far fewer people intimately involved in elephant networks. At those earlier times Asian empires amassed vast armies of elephants, some numbering in the thousands. These animals required constant care, feeding, and training, employing huge numbers of caretakers and veterinarians. In India, elephants were captured from forests, using trained elephants to drive them into corrals (Trautmann 1982). In Southeast Asia, elephants were key workers in the logging industry. Today, trained elephants are far fewer in number throughout Asia than there were in the past (Sukumar 2006), but now hundreds of people research wild elephants in Asia and in Africa, studying their sociality and forms of communication, diet and disease, and their effect on the flora of their habitats. Elephant sanctuaries and orphanages are being built or expanded in Asia and Africa, and in the latter, hundreds of armed rangers work under 'shoot to kill' orders to stop poachers who seek elephants for their valuable ivory. These elephants have bodyguards, and even military helicopters are enlisted in the struggle to save elephants. Other groups take a more peaceful approach, working in an emerging industry that attempts to mediate difficulties between wild elephants and humans, known as 'human–elephant conflict'. Researchers studying human–elephant conflict, especially in Africa, have conducted experiments using techniques such as building steep ditches and walls around farmers' fields. However, they struggle to find methods that are affordable and effective. Some of the more successful techniques have been imported to Asia where, in Malaysia for instance, elephants can destroy heavily capitalised oil palm plantations. In those places, with millions of dollars to lose, so-called elephant-proof fences have been constructed, but these often fail to keep out the elephants. Full-time guards are hired to regularly patrol and to repair the inevitable damage that elephants, and fallen trees and heavy storms, cause to the fences.

In China, officials in Yunnan do not have this kind of funding for robust fences and permanent guards, but they have been experimenting with paying villagers to create what they call 'dinner halls' for elephants. In this plan, villagers grow bananas, corn and rice for the elephants to eat, with the hope that this will deter elephants from raiding the villagers' own crops. Yet the villagers remain sceptical of its effectiveness. As I was told:

> When is an elephant full? They just eat and eat all day long – if there is food around, they might not stop. They will eat from the officials' field, then they will come for mine. (field notes, 12 March 2001)

The International Fund for Animal Welfare has also experimented with 'crop substitution' in Yunnan, trying to convince farmers to abandon crops like rice, corn, and sugar cane that elephants desire and seek out and replace them with crops avoided by elephants, such as chilli peppers. In Africa, scientists are trying to understand elephants' ability to produce infrasonic sound and to 'hear' with their feet, in part to find non-violent ways to keep elephants away from farmers' fields (Shwartz 2005). The use of such sounds would be quite complicated within regions, as in China, where humans and elephants live together in relatively high densities, in complex mosaics, not in two distinct spaces. In Yunnan Province there are many thousands of fields tended by relatively poor farmers, interspersed with forest and other elephant thoroughfares, thus creating a massive logistical and financial impediment. None of these technologies are totally successful: elephant-proof fences need constant maintenance, birth control is expensive and difficult to implement, and the use of infrasonic sounds to keep elephants away is still in its early stages. Further, crop substitution does not work so well in many places where farmers mainly grow for their own food supply, so it is difficult for farmers to switch from rice or corn to chilli peppers.

In viewing elephants as agents instead of as victims of human activities we might understand their behaviour not merely as some kind of inherited instinct, driving them to endlessly repeat old patterns, or as a form of resistance against human activities but rather as an active desire to explore new places, to interact with and to learn about new technologies such as electric fences, to seek out particularly loved foods and to experiment with new ones, and to learn their way around new infrastructure like a major highway. These forms of elephant learning likely play an important role in how species networks are changing in China, and the different ways in which elephants adapt to new scenarios.

A villager in China told me that the elephants of the Mao Zedong era differed from the elephants of today. These live in a world where they are not hunted. Before I heard that, I had never really thought about non-human animals as historical actors whose behaviour as a group can change relatively quickly. I had only heard of one case of evolutionary adaptation that happened within the span of a single human lifetime. It is the famous instance of a moth in England that changed during the Industrial Revolution. As the levels of soot increased, the percentage of darker-coloured moths gradually outnumbered those of lighter-coloured moths, as the darker ones had better camouflage and they were more likely survive and to propagate. Environmental historians had begun to include non-humans in their accounts, but there was no real sense that the behaviour of the animals were changing, as we might expect human society to change (Anderson 2006; Ritvo 1987). With this perspective we let go of the sense that non-humans exist purely in the realm of fixed instinct and instead understand them as fellow learning creatures, figuring things out individually and also sharing these insights with their peers. It opens up scholarship to be attentive to these kinds of dynamic social change for non-humans and humans that expands the realm of what has been called agency.

## Conclusion

The wild elephants of Yunnan Province are actively involved in a place-making deeply shaped by entanglements with humans. To only look at their forms of 'resistance', whether this is raiding crops or walking on highways, is to miss so much of elephant activities. Such a stance continually places humans back in the centre: non-humans only exhibit agency to the degree that they challenge human intentions. This is part of what can be wrong-minded about notions of the Anthropocene that reinvigorate a deep sense of anthropocentrism, part of a longstanding tendency in Western thought that is certainly part of what had led us to such a dire state of the environment today. Seeing other species only in relationship to ourselves, and as largely passive automatons incapable of thinking, learning and feeling, is part of what had led humans to a more disenchanted sense of the world, and give short shrift to what it might mean to 'live well with others'. The other possibility – of seeing non-humans only as victims is also problematic – for it strips these beings of their own actions and their own sensibilities. Looking instead at non-humans as active beings and historical actors that shape the worlds that both they and other species – including humans – live in, can help us attend to seeing ourselves as co-participants with non-human species, imagining new worlds that eschew anthropocentrism. Here, we can see these emerging networks around elephants in China not only how they expand but also how they are reconfigured over time, in part by the actions of the elephants themselves.

## Works cited

Anderson VDJ (2006). *Creatures of empire: how domestic animals transformed early America.* New York: Oxford University Press.

Bates LA, Sayialel KN, Njiraini NW, Moss CJ, Poole JH & Byrne RW (2007). Elephants classify human ethnic groups by odor and garment color. *Current Biology* 17(22): 1938–42.

Bennett J (2007). Edible matter. *New Left Review* 45: 133–45.

Callon M (1986). Some elements of a sociology of translation: domestication of the scallops and the fisherman of St Brieuc Bay. In J Law (Ed). *Power, action and belief: a new sociology of knowledge* (pp196–223). London & Boston, MA: Routledge & Kegan Paul.

Crutzen P & Stoermer EF (2000). The 'Anthropocene'. *Global Change Newsletter* 41:17–18. Retrieved on 29 April 2015 from http://www.igbp.net/download/18.316f18321323470177580001401/NL41.pdf.

Dibley B (2012). 'The shape of things to come': seven theses on the Anthropocene and attachment. *Australian Humanities Review* 52: 139–53.

Elvin M (2004). *The retreat of the elephants: an environmental history of China.* New Haven, CT: Yale University Press.

Foucault M (2009). *Security, territory, population: lectures at the College de France 1977–1978.* London: Macmillan.

Hathaway MJ (2013). *Environmental winds: making the global in Southwest China.* Berkeley, CA: University of California Press.

Hribal J (2007). Animals, agency, and class: writing the history of animals from below. *Human Ecology Review* 14(1): 101–12.

Ingold T (2011). *Being alive: essays on movement, knowledge and description.* London: Routledge.

Latour B (2004). *Politics of nature: how to bring the sciences into democracy.* Cambridge, MA: Harvard University Press.

McFarland SE & Hediger R (Eds.) (2009). *Animals and agency: an interdisciplinary exploration.* Boston, MA: Brill.

Marx, K (1970 [1845]) *The German ideology.* New York: International Publishers.

Palmer C (2006). Killing animals in animal shelters. The Animal Studies Group (Eds). *Killing animals* (pp170–87). Champaign, IL: University of Illinois Press.

Pan WJ, Lin L, Luo AD, & Zhang L (2009). Corridor use by Asian elephants. *Integrative Zoology* 4(2): 220–31.

Philo C & Wilbert C (Eds) (2000). *Animal spaces, beastly places.* London: Routledge.

Prestholdt J (2007). *Domesticating the world: African consumerism and the genealogies of globalization.* Berkeley, CA: University of California Press.

Ritvo H (1987). *The animal estate: the English and other creatures in the Victorian Age.* Cambridge, MA: Harvard University Press.

Robbins LE (2002). *Elephant slaves and pampered parrots: exotic animals in eighteenth-century Paris.* Baltimore, MD: Johns Hopkins University Press.

Rothfels N (2002). *Savages and beasts: the birth of the modern zoo.* Baltimore, MD: Johns Hopkins University Press.

Scott JC (1985). *Weapons of the weak: everyday forms of peasant resistance.* New Haven, CT: Yale University Press.

Shwartz M (2005). Secret sounds of elephants: new research suggests that Africa's pachyderms may be using their trunks and feet to 'listen' to various communications vibrating through the earth from other elephants. *National Wildlife*, 1 May. Retrieved on 29 April 2015 from http://www.nwf.org/news-and-magazines/national-wildlife/animals/archives/2005/secret-sounds-of-elephants.aspx.

Stoppani A (2012). A new force, a new element, a new input: Antonio Stoppani's Anthropozoic. V Federighi (Trans), E Turpin (Ed) from the 1873 edition of the *Corso di Geologia*. In E Ellsworth & J Kruse (Eds). *Making the geologic now: responses to material conditions of contemporary life* (pp34–41). Brooklyn, NY: Punctum Books. Retrieved on 29 April 2015 from http://geologicnow.com/2_Turpin+Federighi.php.

Sukumar R (2006). A brief review of the status, distribution and biology of wild Asian elephants *Elephas maximus*. *International Zoo Yearbook* 40(1): 1–8.

Thompson C (2002). When elephants stand for competing philosophies of nature: Amboseli National Park, Kenya. In Smith BH, Weintraub ER, Law J & Mol A (Eds). *Complexities: social studies of knowledge practices* (pp166–90). Durham, NC: Duke University Press.

Trautmann TR (1982). Elephants and the Mauryas. In Mukherjee SN (Ed). *India: history and thought* (pp254–81). Calcutta: Subarnarekha.

Whatmore S & Thorne L (1998). Wild (er) ness: reconfiguring the geographies of wildlife. *Transactions of the Institute of British Geographers* 23(4): 435–54.

# Epilogue
## *New World Order* – nature in the Anthropocene

*Hayden Fowler*

> Probably no society has been so deeply alienated as ours from the community of nature, has viewed the natural world from a greater distance of mind, has lapsed into a murkier comprehension of its connections with the sustaining environment.
>
> Nelson 1993, 202–3

The video *New World Order* (2013) depicts an ashen forest of blackened and petrified trees. Its grey and tangled, barren aesthetics are reminiscent of both the fairytales of a deeper past and the apocalyptic nightmares of the not-too-distant future. Occupying and animating this eerie world is a strange collection of birdlife. Preening high up on branches, scratching the crusted ground or slipping beneath the tree roots into underground burrows, these creatures appear at home in this unfamiliar place. At different times a bird will sing out, emitting a strange electronic call of blips and clicks that punctuates the silence of the dark forest. Presented as a series of sliding vignettes, the video draws on both historical and contemporary depictions of nature, from

Fowler H (2015). Epilogue: *New World Order* – nature in the Anthropocene. In Human Animal Research Network Editorial Collective (Eds). *Animals in the Anthropocene: critical perspectives on non-human futures*. Sydney: Sydney University Press.

landscape paintings and museum dioramas through to wildlife documentary films.

The forest engages the biophilic response of the viewer by employing the cues of natural beauty, while undermining this with suggestions of death and devastation – intimately entwining loss and desire in a reflective or critically nostalgic image. Typical of contemporary utopian imaginings where hope can only survive in proposed futures or sideways paradigms, *New World Order* also exists out of time. Alluding to aspects of both the past and the future, it suggests a new and autonomous nature evolving and recovering just beyond some apocalyptic wake of humanity. The incorporation of animals in the work is a further means of emotionally engaging the viewer, through replicating the profound experiences of encountering and of observing an animal in the wild and of discovering and of exploring new and unfamiliar natures. In the manner of abstract utopias the work pursues the education of desire.

The initial experiences of 'nature' are undermined by a sense of blurring between authentic or synthetic nature and as the video progresses, the viewer is challenged to make sense of what is artifice and what is real. The entire forest is a painted construction of timber, plaster, sticks, cement and dirt. The birds themselves, canaries and heritage-bred chickens, have all been selectively bred by humans over centuries, sometimes millennia. The forest is thus truly anthropocenic on a number of levels. Its physical construction and wildlife are a blurred mix of anthropogenic nature, and the work denies a restorative fantasy through the depiction of nature as depleted and transformed. Its apocalyptic aesthetics visualise a conceivable future, based on the prevailing environmental crises. The desolation tapping into the grief and mourning for a devastated biosphere, but its surviving ecology and viewer emotions can offer hope or possibility for human re-engagement with nature in this transformed new world order.

My art practice explores humanity's relationship with the natural world and the broader historical and cultural concepts that influence this engagement. Incorporating video, photography, installation and performance, my work involves complex productions that include long periods of research, the construction of elaborate sets and the specialised training of domestic animals, such as goats, chickens, lambs, rats and canaries. All these species have a history entwined with civilisation and are encoded with layers of cultural symbolism from paganism to contemporary religion, science, colonialism and capitalism. Recurring themes within my work are of nostalgic and utopian desire, freedom, loss and blocked romantic hopes for a returned and re-enchanted intimacy with nature. I am exploring these ideas as an emerging discourse in contemporary art; that which is capable of critically and of poetically navigating the transformed human–nature relationship within the Anthropocene, through the construction of imagined alternate futures or paradigms.

With the dawning recognition of the Anthropocene, an already alienated humanity finds itself in a transformed and depleted natural world. In the past 40 years the Earth's entire population of wildlife has fallen by 52 percent (World Wildlife Fund for Nature 2014, 8) and within this century 20–70 percent of all of the Earth's surviving species are predicted to be at risk of extinction (Sandler 2012, 76). Jeremy Bendik-Keymer argues that what is largely misunderstood in contemporary environmental discourse is that for millions of years following any mass-extinction event, and therefore within any 'conceivable human future', the Earth will be a 'bio-wasteland', 'eerily empty of biodiversity' and 'pared down to basics' (Bendik-Keymer 2010, 10–12). Sandler affirms that 'our ecological future is accelerating away from our ecological past with increasing rapidity and that it is increasingly unclear where it is going' (Sandler 2012, 65) and in the words of Bill McKibben (2010), we live on a 'new earth' (quoted in Sewall 2012, 270).

While the Anthropocene is largely defined by the depletion, degradation and transformation of the physical world, my art practice explores the emotional, psychological and cultural bonds between humanity and nature, the consequences of this loss and the prevailing human separation from nature. The destruction and transformation of external nature is only one aspect of the contemporary human–nature separation; the processes of civilisation, colonisation, Enlightenment, capitalist-industrial process, and urbanisation of the population have over time, produced what many ecopsychologists contend is an *internal* environmental crisis. Laura Sewall calls it a crisis of 'perception', that humans no longer readily see the patterns and systems that reveal our dependence and interconnection with the natural world (Sewall 265).

Peter Kahn describes this loss of perception as *environmental generational amnesia* that while environmental degradation is increasing with every generation, each new generation generally regards the environment they were born into as the non-degraded norm (Kahn 1999 quoted in Kahn & Hasbach 2012, 319).

Humans, however, are deeply biophilic creatures with an innate tendency to affiliate with life, as Kahn and Hasbach write:

> We need nature for our physical and psychological wellbeing. We always have. Our bodies and minds came of age interacting with a natural world that sustained and resisted our being and through both processes shaped the contours of what it means for humans to flourish. (Kahn & Hasbach 2012, 1)

Sewall points out that with the experiential separation from nature

> we have forgotten the slippery magic of water sliding through toes; rivers running thick and wild with salmon, shad, and alewives; or cedar trunks spanning six feet across, their tree tops sweeping the sky and transporting human imagination towards the heavens. Such a loss of collective, embodied knowledge cannot be understated. It is a forgetting of the perceptive and sensual beings we naturally are, of imaginal reach, and of the sheer throbbing vitality and wholeness of the organic world. (Sewall 2012, 274)

The external degradation of the diversity and health of nature is thus intimately connected with a degradation of the human psyche. We have lost intimate daily connection with nature, resulting in the *extinction of experience* (Pyle 1998 quoted in Sampson 2012, 23–24) and the atrophying of our knowledge and sensory capacities for connecting with and understanding the natural world.

On a cultural and consciousness level, the recent labelling of an Anthropocene and the understanding of its ramifications, mark a significant moment in the transformed human–nature relationship. The ground on which our relationship to nature is built, has shifted, disappeared, become illusory – our previous narratives and discourses relating us to nature have become redundant. While humanity continues to myopically soothe or to deny this separation through nature documentaries, restorative conservation efforts or trips to the zoo – a critical view of nature in the Anthropocene requires a realisation that most of what we will hold on to as nature is a falsehood, virtual and imaginary remembrances of what are largely fragmented and depleted remains. And, as the Anthropocene marks a crisis point in our physical relationship to the natural world, it also signifies a barely recognised ideological, emotional and psychological turning point on how we re-calibrate, re-engage and re-enchant our relationship with a transformed natural world and imagine alternative futures – a task we are ill-equipped to navigate.

Few modes of language for acknowledging and for grieving these losses, or for reconnecting with nature, are currently recognised. This is, in part due to reflective, emotional or poetic human–nature discourses being eliminated by the prevailing ideologies of rationalist modernity. In my practice and research I have been working with four discourse structures: Romanticism, Utopianism, Nostalgia and Mourning. These four have evolved as alternatives to the prevailing rationalist paradigm. In my work, the amalgamation of these four alternatives form a critical–poetic framework for examining human–nature engagement, and for imagining other ways forward. Romanticism's history as a counter to the Enlightenment's rationalist dominance continues to have relevance in contemporary environmental discourse, particularly because of its understandings of ecological interconnectivity, its philosophies of relating to nature as an equal other such as Novalis' 'Nature as a *You*', and as a project to re-enchant nature in the human imagination. Utopianism, in its abstract form, allows for the imagination of alternative paradigms, embodying hope (or fear in the case of dystopias), educating desire, and inspiring the pursuit of a transformed world. Nostalgia, defined as the 'longing for a home that no longer exists or never existed' (Boym 2001, xiii), poetically articulates the experience of nature in the Anthropocene – and in its critical reflective form it understands the impossibility of return, but draws on, and attempts to heal, the past in its imagination or construction of new futures. Finally, the language of mourning re-evaluates nature and the non-human as grievable bodies (Butler 2009, 39), generating human–nature equality through an experience of shared vulnerability, and makes public what currently exists as private, unspoken experiences of grief and the loss and degradation of nature around us.

## Works cited

Bendik-Keymer J (2010). A conceivable human future: time & morality in the sixth mass extinction. A lecture at the Baker-Nord Centre, Case Western Reserve University, Cleveland, OH, 8 April.

Boym S (2001). *The future of nostalgia*. New York: Basic Books.

Butler J (2009). *Frames of war: when is life grievable*. London: Verso.

Fowler H (2013). *New World Order*. Video artwork, Sydney, Australia. Available online at http://haydenfowler.net/projects/new-world-order.html.

Kahn PH (1999). *The human relationship with nature: development and culture*. Cambridge, MA: MIT Press.

Kahn PH & Hasbach PH (Eds) (2012). *Ecopsychology: science, totems, and the technological species*. Cambridge, MA, London: MIT Press.

McKibben B (2010). *Earth: making a life on a tough new planet*. New York: Times Books.

Nelson R (1993). Searching for the lost arrow: physical and spiritual ecology in the hunter's world. In SR Kellet & EO Wilson (Eds). *The biophilia hypothesis* (pp201–28). Washington, DC: Island Press.

Pyle RM (1998). *The thunder tree: lessons from an urban wildland*. Guilford, CT: Lyons Press.

Sampson SD (2012). The topophillia hypothesis. In PH Kahn & PH Hasbach (Eds). *Ecopsychology: science, totems, and the technological species* (pp23–53). Cambridge, MA & London: MIT Press.

Sandler R (2012). Global warming and virtues of ecological restoration. In A Thompson & J Bendik-Keymer (Eds). *Ethical adaptation to climate change human virtues of the future* (pp63–79). Cambridge, MA: MIT Press.

Sewall L (2012). Beauty and the brain. In PH Kahn & PH Hasbach (Eds). *Ecopsychology: science, totems, and the technological species* (pp265–84). Cambridge, MA & London: MIT Press.

World Wildlife Fund for Nature (2014). *Living planet report*. Gland, Switzerland: World Wildlife Fund for Nature.

*New World Order* (2013) video stills by Hayden Fowler.
HD digital video, colour, sound 15 min 17 secs.
Available at: http://haydenfowler.net/projects/new-world-order.html

# About the contributors

**Vanessa Barbay** is a practising artist from Vincentia, a village on Jervis Bay, New South Wales. Barbay originally left Vincentia to pursue an education in visual art at Charles Sturt University in Wagga Wagga. After a transient period she returned to the Riverina and worked as a tattooist, mural artist, and photographer at Kapooka Army Base. She also taught painting, drawing and art history at TAFE. After moving to Canberra, she undertook postgraduate research at the Australian National University (ANU). Barbay recently completed a postdoctoral painting project in collaboration with Koori artist Theresa Ardler through the ANU Vice-Chancellor's Visiting Artists Scheme. She exhibits regularly and has been published in *Antennae* and *Art Monthly*. Barbay has returned to live in Vincentia and is currently undertaking a five-week residency at Bundanon Trust where she is working toward a joint exhibition at the Shoalhaven City Art Centre in June 2015. Her postgraduate research and current practice can be viewed at http://laomedia.com/blog.

**Gwendolyn Blue** is an assistant professor in the Department of Geography at the University of Calgary, Canada. Formally trained in cultural studies, her research interests centre on public engagement with environmental and health issues. Current research covers two domains: public participation in environmental governance, with a focus on deliberative democracy and climate change, and emergent manifestations of the public in contexts in which the boundaries between humans, animals and technologies have been called into question.

**Madeleine Boyd** is driven by a series of intense inquiries that involve thinking with non-human animals and the matter of existence. Currently engaged in a process of discovering what it is like to 'intra-act with horses', she presents her findings as a series of public videos, online blogs, art installations and paddock-based happenings. Madeleine is in the final stages of a PhD candidature in Sculpture, Performance and Installation at Sydney College of the Arts, University of Sydney. She co-curated issue 31 of *Antennae: Journal of Nature in Visual Culture* featuring a selection of papers on multispecies art practices as they intersect Karen Barad's agential realism. Madeleine generally feels that humans are a dangerous, carnivorous species to be avoided whenever possible, particularly if you are a horse.

**Florence Chiew** is currently the Higher Degree Research Learning Advisor in the Faculty of Human Sciences at Macquarie University, Sydney. Her research is driven by the 'two cultures' problem that asks how different notions of truth, objectivity and scale can be reconciled across the sciences and the humanities. Florence's recent project, 'Systematicity: the human as ecology', explores how foundational questions in the social sciences about individual agency anticipate many of the contemporary debates on the ethics and politics of human–non-human relations. This project is an analysis of how we conceive our place in nature as broader ecological concerns redefine and challenge what we mean by nature and what it means to be human.

**Matthew Chrulew** is a Research Fellow in the Centre for Culture and Technology at Curtin University, Perth. His essays have appeared in *Angelaki, SubStance, Environmental Humanities, New Formations, Foucault Studies, Australian Humanities Review, Humanimalia, Antennae, The Bible and Critical Theory*, and the collections *Animal Death* and *Metamorphoses of the Zoo*. He is an associate editor of the journal *Environmental Humanities*. With Chris Danta he edited issue 43(2) of *SubStance* on Jacques Derrida's lectures on *The Beast & the Sovereign*, and with Jeffrey Bussolini and Brett Buchanan he edited issues 19(3) and 20(2) of *Angelaki* on the philosophical ethology of Dominique Lestel and Vinciane Despretl.

**Chris Degeling** is a Research Fellow at the Centre for Values, Ethics and the Law in Medicine at the University of Sydney. A practising veterinarian, his research and teaching interests revolve around the ethics and politics of human interactions with non-human animals, and the social and cultural dimensions of public health. His research is interdisciplinary, and draws together insights from Science and Technology Studies (STS) and normative theories. Current projects include studies of social justice and pet ownership, the politics of 'One Health', and the ethics of cancer screening.

**Simone Dennis** is senior lecturer in anthropology at the Australian National University, Canberra. Her research interests coalesce around anthropological theories of embodiment, power and the senses. Dennis' ethnographic work on Australian research laboratories, in which rodents feature as animal models for human disease research looks at how scientific ideas about kinship were applied in the context of posthuman biopolitics that pervaded the laboratories in which she worked. It also reworks anthropological notions of kinship to better understand the traffic across the human–animal divide. She is the author of three monographs: *Police beat: the emotional power of music in police work* (2007), *Christmas Island: an anthropological study* (2008) and *For the love of lab rats: kinship, humanimal relationships and good scientific research* (2011). She appears on Australian national television and radio programs and speaks at public forums to communicate the findings of her anthropological work to the broader community.

**Ben Dibley** is a research fellow at the Institute for Culture and Society, the University of Western Sydney. His research interests are in social and cultural theory, particularly on questions of colonialism and the environment. He has recent publications in *Australian Humanities Review*, *History and Anthropology*, *New Formations*, and *Transformations*.

**Hayden Fowler** is a New Zealand born artist, based in Sydney, Australia. He holds a Master of Fine Arts degree from UNSW Art and Design (COFA) in Sydney, as well as an earlier degree in biology. Fowler's methodology involves the development of elaborate set constructions in which he choreographs human and animal subjects, creating hyper-real video, photographic, installation and performance work from within these fictional spaces. His practice explores the unsettled human relationship with the natural world in the emerging Anthropocene, drawing on the historical conditions that have influenced this engagement. Fowler has exhibited nationally and internationally, and his work is held in a number of public and private collections. He is a previous recipient of the Samstag International Visual Arts Scholarship, undertaking a year of study abroad at the Universitat der Kunst in Berlin, Germany. He lectures in the Sculpture, Performance and Installation studio at UNSW Art and Design.

**Adrian Franklin** is professor of sociology at the University of Tasmania, Hobart. Adrian has been a contributor to the field of human–animal relationships since its early days, with *Animals and modern culture* (1999), followed by *Nature and social theory* (2002) and *Animal nation* (2006). His work on human–animal relations focuses on the shifts brought about by modernity; changes in perceptions of nature, the advent of posthumanism and, in *Animal nation*, questions of nationalism, belonging and identity. His most recent book in this area, *City life* (2010) challenges the concept of the city as a humanist citadel, purified of other life. His recent research has investigated species cleansing in Australia, the mangled science of feral animals, the benefits of companion species to human health and loneliness and the big cat phenomenon in the UK.

**Michael Hathaway** is associate professor of cultural anthropology at Simon Fraser University in Vancouver, British Columbia. His first book, *Environmental winds: making the global in Southwest China* (2013), explores how environmentalism was refashioned in China, not only by conservationists but by rural villagers and even animals, including wild elephants. His second major project examines the global commodity chain of the matsutake, one of the world's most expensive mushrooms, following it from the highlands of the Tibetan Plateau to the markets of urban Japan. He works with other members of the Matsutake Worlds Research Group, looking at the social worlds this mushroom economy engenders in Canada, the USA, China, and Japan.

**Daniel Kirjner** is a vegan–feminist activist and doctoral sociology student at University of Brasília, Brazil. His research focuses on links between the construction of masculinity and the glorification of violence against animals and women in contemporary capitalist societies. Since 2011, Daniel has offered his presentation, 'An invisible veil: a feminist look at animals' destinies' at Brazilian universities and social collectives. Daniel has had articles accepted for publication in journals and anthologies in Australia, Brazil and the USA. He continues to learn about, question, and deconstruct masculinity in his own life in order to heighten his own awareness of predatory male behaviour in the hope of being himself part of bringing change to the world.

**Agata Mrva-Montoya** has been a member of the Human Animal Research Network at the University of Sydney since 2011. Her PhD focuses on 'cultural faunas' of ancient Cyprus, reflecting the changing perception of various animal species in the context of larger historical shifts in the Cypriot society. Her research focuses on the relationship between symbolism, attitude to, and treatment of, animals, and the ethnic makeup of people in ancient and modern Cyprus.

**Richie Nimmo** is lecturer in sociology at the University of Manchester in the UK, where he teaches human–animal relations and environmental sociology. His research is interdisciplinary in nature, and involves exploring the ambiguous status of non-humans in modern knowledge practices and the constitution of the social across materially heterogeneous relations, systems and flows. This interest is both historical and contemporary, and he has published in journals including *Society and Animals* and *Journal of Historical Sociology*. His first book, *Milk, modernity and the making of the human: purifying the social* (2010), was a socio-material history of dairy milk in the UK. Currently he is thinking about bees, pesticides and colony collapse disorder, and working on an edited collection on actor–network methodologies.

**Fiona Probyn-Rapsey** is senior lecturer in the Department of Gender and Cultural Studies, University of Sydney. Since 2011 she has been convenor (now co-convenor with Dinesh Wadiwel) of the Human Animal Research Network at the University of Sydney. Fiona is vice-chair of the Australasian Animal Studies Association (previously known as the Australian Animal Studies Group) and is on the editorial boards of *Environmental Humanities, Australian Humanities Review* and *Animal Studies Journal*. She is author of *Made to matter* (2013), co-editor (with Jay Johnston) of *Animal death* (2013). She is also co-editor (with Melissa Boyde) of the Animal Publics series at Sydney University Press.

**Nikki Savvides** completed her PhD in cultural studies at the University of Sydney. Her thesis analyses the work of a number of conservation and animal welfare projects at various sites across South and Southeast Asia. She has published research papers on human–horse relationships and a study on Bangkok's stray ('soi') dogs that have appeared in *Society and Animals, Humanimalia* and the *Animal Studies Journal*. Nikki is currently conducting an ongoing ethnographic study of the complex relationships between mahouts and their elephants in a small tribal village in northeast Thailand.

**Krithika Srinivasan** is a lecturer in human geography at the University of Exeter, the UK. Her research and teaching span animal studies, more-than-human geographies, environmental politics, and alternative development. Particular interests include animal welfare and conservation politics, urban more-than-human inclusivity, compassionate conservation, and cross-cultural investigations into environmental and animal justice. Krithika is from India where she studied and worked at the Tata Institute of Social Sciences (TISS), Mumbai. She brings the deeply political orientation to research and analysis cultivated during her time at TISS to all aspects of her current engagements with nature-society and more-than-human geographies.

**Dinesh Joseph Wadiwel** is a lecturer in human rights and socio-legal studies at the University of Sydney. His research interests include sovereignty and the nature of rights, violence, race, and critical animal studies. He is author of *The war against animals* (2015).

**Alison Witchard** is a PhD candidate in anthropology at the Australian National University, Canberra. Her previous works include an exploration of the emergent biotechnology of in-vitro meat. Alison is currently undertaking a Fulbright Postgraduate Scholarship as a Visiting Fellow at Harvard University where she is researching the embodied experiences of women at risk of hereditary breast and ovarian cancer.

# Index

# Index

Zalasiewicz, Jan viii, 2, 22
zoopolis xix

Žižek, Slavoj 29